MW01256793

INVENTING SECRETS REVEALED

BRIAN FRIED

WingSpan Press

Copyright © 2016 by Brian Fried

All rights reserved.

No part of this book may be reproduced or transmitted in any form or by any means, electronic or mechanical, including photocopying, recording or by any information storage and retrieval system, without written permission from the author, except for the inclusion of brief quotations in review.

Published in the United States and the United Kingdom
by WingSpan Press, Livermore, CA

The WingSpan name, logo and colophon are the trademarks
of WingSpan Publishing.

ISBN 978-1-59594-573-0 (pbk.)
ISBN 978-1-59594-905-9 (ebk.)

First edition 2016

Printed in the United States of America

www.wingspanpress.com

Library of Congress Control Number: 2015957832

1 2 3 4 5 6 7 8 9 10

PREFACE

A curious thing happens when you write and publish a book. Almost before you know it, people start asking when you're going to write the next one.

If writing books is what you do full time, the answer you're looking for is easy.

"I'm working on it right now," you say with a smile.

If you're the type who likes to plan ahead, you may add,

"Would you like me to add your name to the list of people who already want an advance signed copy?"

It's a great way to grow your pre-qualified contact list, because they already bought your first book and why wouldn't they want your second one? And besides that, how many people are going to say no, after they were the ones who brought the subject up in the first place?

Of course, if you do something else full time to make sure the bills are paid, your answer is going to be a little more complicated. In the first place, there's the exhaustion.

What none of the experts mention when you sit down to write your first book in your spare time is that writing is a lot like building a pyramid. You chip away at it for what seems like an eternity, earning blisters and getting sand in your eyes and- -okay, maybe not as many blisters and almost none of the sand, unless you're writing at the beach, in which case what are you complaining about? Do you have any idea what I would give to be able to write at the beach?

Sorry. I got side-tracked for a second there. But it's all good, because it brings me to the second reason that your "next book" answer can get a little complicated.

So my advice is to do what I've done. Go with the first answer and keep smiling through your tears. Because with each book it gets a little easier.

Okay, no. One little white lie is fine, but that's too much. No book is easy if you put your heart into it. Every single one of them is a labor of love.

Which is why it took me so long to write this one. But here it is, at long last! And yes, my next book is in the works! Thanks for asking. Would you like me to add your name to the people who already want an advance signed copy?

CONTENTS

INTRODUCTION

When I sat down to write *"You & Your Big Ideas"* back in 2007, I had been living a double life for four long years. By day, I was the family bread-winner. By night, when the family was asleep, I was an inventor.

My biggest challenge, back in those days, wasn't coming up with ideas. I'm an inventor. I have ideas all the time.

What set me back, time after time, was the lack of support. Even though I lived in Suffolk County, New York, where American innovations in medical science, aviation and aerospace got their start, my area was extremely limited when it came to resources for the small inventor like me.

If you've followed my career since then, you know that I decided to stop with the grumbling already and do something to change things for people like me.

It took the better part of a year to get the ball rolling, but the effort was worth it when Suffolk County agreed to sponsor the not-for-profit Inventors and Entrepreneurs Club of Suffolk County where inventors go to brainstorm, network and hear experts share their stories about their successes and pitfalls they need to watch out for.

Because I've had some level of success as an inventor – at the time, I'd already patented six products and filed over a hundred trademarks – people started asking me to give talks.

Before I knew it, I'd gathered a lot of material. I didn't want to be known as one of those guys who use the same dusty old PowerPoint presentation every time. Because I was passionate about inventing, I really wanted to keep my presentations fresh and interesting for anyone just getting their feet wet and navigating through the invention process.

After these talks, people started saying I should write a book. When enough of them said it enough times, I said to myself, Why not? So I called up this editor friend of mine and said,

"I've got all this great material. All you need to do is stitch it together. Can you make it a priority and get it done in a month?"

When she'd finally stopped laughing so hard that she couldn't speak and she'd managed to wipe up most of the coffee she'd spilled all over her keyboard, she took me gently by the hand.

By 2008, almost a year later, the work was done. The book was finally in my hands. It was an incredible feeling. I'd invented a book, "You & Your Big Ideas!"

Then I took it to presentations and someone said,
"When are you going to write another?"
Soon is what I told them, but I had other plans. What I'd learned as I wrote was that inventors all over America were looking for help. When I found out that experts in every

aspect of the field were eager to step up with advice and support, the light bulbs began to go off.

I realized that with my connections, I could create the support network they were looking for. Even better, if I used the internet, I'd be able to reach a much wider audience than Suffolk County, New York.

In 2009, I launched Got Invention Radio – a weekly online chat show that allowed me to broadcast live interviews with successful, high profile inventors and resource experts in the field.

We would broadcast live every Thursday evening at 8 PM Eastern Time. Because the show airs on the internet, listeners can access the past interviews at their convenience and listen to their favorites an unlimited number of times. [www.gotinvention.com]

In 2012, continuing to reach out further to the inventor community, several initiatives were on my agenda. I was ready to launch my inventor consulting practice.

InventorConsulting.com is a website that talks briefly about me, offers a free Non Disclosure Agreement (NDA) and connects to my schedule. I offer inventors and entrepreneurs to speak with me and to discuss where they are with their idea and how I can help them.[www.inventorconsulting.com]

GetInventionHelp.com allows inventors to select what step of the invention process they need help with and I connect them to reputable inventor resources. [www.getinventionhelp.com]

I also approached Long Island's neighboring county, to the existing Suffolk Inventors & Entrepreneurs Club and launched Nassau County Inventors & Entrepreneurs Club. Founding and running both clubs covered the entire Long Island inventor community, which if both counties of Long Island combined, it could make up the 17th largest state in the US.
Suffolk I&E Club [www.meetup.com/iesuffolk] Nassau I&E Club [www.meetup.com/ienassau]

My inventingtips.com blog also casted a following and I made very good use of my time commuting to Manhatten daily on the trains and subways, writing about a variety of topics and how to overcome inventor and inventing challenges.[www.inventingtips.com]

In 2013, the people at **answers.com** chose me as their Content Expert Writer for the "INVENT" category. Each month, my articles help inventors with new ideas that are going through the invention process. I share inventing tips, offer various inventor resources, and let them know about trends and current events within the inventor community.
[http://www.answers.com/brianfried]

In the 4th quarter of 2015, in between editing of this book, I am preparing for the start of my new brand, Inventor Smart, will license products and serve as an aggregator for inventors with their inventions turned products, have inventory and need distribution under one brand to present to retailers. Inventor Smart will also be an e-commerce platform for inventors products where customers can purchase the latest and greatest inventions and learn more about the inventor that created it. It will also become the new hub for all my websites, including the creation of the Got Invention? resource center with blog posts, radio show interviews, book recommendations and providing services at each stage of the developing an invention. [www.inventorsmart.com]

In my spare time, I kept working on my own inventions. At the time of this writing, I have many products of my own inventions and other inventors products in retail distribution channels, including As Seen On TV, Home Shopping channels, mass retailers, online and through catalogs, some by licensing and others I have helped to manufacture.
I recently received my Knot Out, Knot Loosening Device patent ready to launch. Also, after many years of challenging the US Patent Office, received my Pull Ties patent now issued. I have some more inventions on the way!

What with all that going on, and a busy family life and a full-time job, finding the time to focus on book number two always stayed just out of reach. Until now…

Inventing Secrets Revealed was written to satisfy three goals. The first – to finally answer the question I began hearing back in 2008. I know, the next question will soon be coming, "when is book number three" and I have no one but myself to blame!

The second – to offer up the insights I've learned since the early days. In any industry, it's important to stay on top of the trends that change the way things are done.

And the third – to share my stories as an inventor, partly to illustrate with real life examples the information I provide, and partly as a way to help anyone who may be wondering if the bumps in the road only happen to them. Trust me, it's not just you. We're all in this together!

And now, let's get started…

PS Congratulations on following your dream!

INSPIRATION TO DREAM...

The world is filled with people who pursued dreams that other people told them they would never bring to life. They may have been told that their dreams were foolish, or risky, or that the odds were stacked against them. Some of those people tried anyway and failed repeatedly before they succeeded. The difference between the critics and the people who succeeded in bringing supposedly unrealistic dreams to reality is that they continued on. They persevered. They didn't listen to what others said - or they listened and rejected what they heard. Keep your eye on your dream, and keep going. Persistence is a priceless element of success.

– Copyright (c) Daily Horoscope

FIRST STEPS

Sometimes it seems impossible. There are set-backs. Failures. Unexpected costs. There are dishonest people and people who criticize your every move. I know how you feel. Every dreamer like you has stood where you're standing now. Even the really great moments carry with them a small seedling of doubt. Here's the secret… It's not your challenge to overcome all this stuff. It's your challenge to keep on moving forward, no matter what.

– Brian Fried

This section of *Inventing Secrets Revealed* almost wrote itself. The people I need to thank for the inspiration includes anyone who has had the courage to take those first hesitant steps and then – when they lost their way or something went wrong – they dug in their heels and made the brave decision to ask for help. When they got it, the next thing they did after that was keep on going.

Are You An Inventor? Let's Check Your Profile

Are you already an inventor or do you want to be an inventor? What does an inventor act or think like? When you call yourself an inventor, it can take on a whole new meaning to your life. You can live up to the title by being aware of how things work in deeper detail and this thought process can become second nature to you.

According to the Merriam-Webster Dictionary, the definition of an inventor is someone who, "creates or produces (something useful) for the first time".

As any successful inventor will tell you, the research you do before you set out on your journey can help bring you closer to your goals. So let's evaluate "you" as an inventor by answering the questions below.

<u>Do you have an inventive mind?</u>
Do you come up with ideas and solutions to challenges on a daily basis? Inventive minds are always seeing the next logical step that could be taken, while the rest of the world simply shrugs its shoulders and makes do with things as they are.

Do you say, "This could be done differently" or "that could be improved", "this could be modified a bit" or "that could be a big winner"? Inventive minds get excited when there is a puzzle to solve. Do you see opportunity everywhere you look? Inventive minds are positive and pro-active. They consistently ask themselves, "Why not?" when other people give up.

Do you have inventive habits?

1. Are you curious?
2. Are you the kind of person who asks questions?
3. Do you find yourself wondering how things work?
4. Do you take things apart and see if they'll go back together in a different, more efficient way?
5. Do you make notes and sketches and conduct experiments?
6. Do you like to keep track of your ideas and use notes and sketches to fix and tweak them from time to time?
7. Do you brainstorm with others and get them excited by your ideas?
8. Is it as much fun to talk about the process as it is to work on it?
9. Do you find yourself feeling energized when you share your ideas and get input from others?
10. Are working on some idea that you're nearly ready to talk about?
11. Have you done your due diligence?
12. Have you researched the market and compared your concept with others currently on the shelves?

13. Have you registered your patent to protect your concept?
14. Are you at the prototyping stage?
15. Have you created a sample that allows other people to see and touch your idea and believe in its possibilities as you do?
16. Do you have a design engineer who is willing and able to take your prototype and create a professional design?
17. Have you looked for opportunities to partner with a designer and take advantage of his established contacts?
18. Are you ready to go to the marketplace?
19. Have you decided whether you want to manufacture it yourself or license it to someone else and let them pay you royalties?
20. Have you registered your business and surrounded yourself with the best team to help you reach your goals?

<u>Do you have a goal?</u>

1. Do you want to solve problems and overcome challenges?
2. Are you driven by a desire to make the world a better place?
3. Are you searching for opportunities to beat the odds and win?
4. Do you want the personal satisfaction of a job well done?
5. Are you the kind of person who sets high goals and standards and consistently meets and surpasses them?

6. Do you want to be famous?
7. Are you looking for the recognition that comes with success?
8. Do you want to be rich?
9. Are you focused on the tangible rewards that come with high volume sales and repeat business?

Do you have what it takes to achieve success?

To take a great idea or concept from brain wave to prototype to marketplace, it takes more than dedication, determination and drive. To do your due diligence, to research and refine, to educate and inform, to overcome setbacks, to protect your work, to form partnerships and leverage your success takes more than perseverance and people skills.

Whether you're inventive and imaginative; whether you brainstorm or work alone; whether you're still thinking about taking that first step or you're already working on your big idea, the one quality you have to have in order to succeed as an inventor is patience with the process.

10 Steps from Idea to Invention

1. First Steps. It all starts with an idea in your head. Wherever you are, you need to get it out of your head. Use a napkin, text, email or leave yourself a voice message to capture your idea before it disappears!

2. Protection. You need to arrange for intellectual property protection, including provisional, design, non-provisional or utility patent and/or trademark.

3. Development. Bring your product to life with computer drawings (CAD), identify materials needed and make a prototype.

4. Licensing. Prepare a list of licensees and representation and show your invention's intellectual property is for them and their customers.

5. Manufacturing. This is when you prepare for production and tooling quotes, broker representation.

6. Marketing. You need packaging, sales sheets, demo videos, preparing the pitch to retailers, a website and social media presence.

7. Distribution. Choosing distribution channels- box big retailers, catalog, as seen on TV, specialty retailers, online retailers, etc.

8. Media. To increase your opportunities for your invention, you need to get some press!

9. Have fun. Inventing is a lot of work. Be sure to take the time to enjoy the process. Remember that you need to make your invention pay its way by earning $$!

10. Get another idea and start all over again!

Start Here With Your Great Idea

Here are some quick tips to get you started on your invention idea. Following these steps will help you to be prepared when you have an idea, evaluate the window of opportunity and focus on the ideas that could have the best chance for success.

Keep That Idea Close

Capture that idea immediately by writing it down, sending yourself a text message, an email or leaving yourself a voice message describing your idea. How many times do you find yourself in the middle of something and you get distracted, maybe the phone rings, someone calls your name and you lose your train of thought only to forget what you were just thinking about.

Visualize Your Idea

Draw what you envision your idea looks to the best of your artistic ability and think about how it would be made. Use a simple piece of paper, sketch it on the computer, your tablet and start to think about what material you could use such as plastic, wood, recyclable material or metals.

Your Idea Used By Others

You may be emotionally attached to your idea, however you should think about how others would feel about it and if they would use it. Why would someone buy it? What do you think it would cost in retail? Who is your target audience? Is your idea for the masses or limited to a select group of potential clients or customers?

Is it Your Idea or Someone Else's?

Perform a search to reveal if your invention idea already exists by check online stores, going to local boutiques, watching home shopping channels during your product category airings, looking through popular print or online catalogs, visiting specialty or big box brick and mortar retailers where you think your product may be sold. Also search online by typing in various descriptions of your idea on several search engines such as Google, Yahoo and Bing, which will show you search results of websites and images to review. Also visit the [U.S. Patent and Trademark Office] website (http://www.uspto.gov) or [Google Patents] (www.google.com/patents), enter various titles to describe your idea in the search field and look at the drawings and descriptions on existing patent results. Start by looking at the artwork and then review the patent holder's claims to their invention.

Track Your Facts

Keep a log of what you find by jotting down the stores that had similar or the same type of products, the brand names, patent numbers, was that and your thoughts to compare and research relative to your idea.

Moving Forward or Moving On With Your Idea

Did you find your idea out in the market already? Is there a big enough difference from what you were thinking to what already exists? Did you not find anything at all that looks like or describes your idea?

Claiming Your Idea

The answers to the questions above may require professional help by a patent attorney or agent for them to read through the patent records and review the products you found. You can find a list of registered intellectual property services in your area on the [U.S. Patent and Trademark Office] (http://www.uspto.gov/) website mentioned earlier. This step may cost you a few dollars but may be well worth it. Imagine skipping over this step, starting on your idea and then finding out that you are infringing on someone else's patent. You should ask for a letter from the lawyer or agent that will give their opinion if your idea may have a chance for intellectual property protection with a patent or other form of protection.

Coming up with your idea, evaluating the opportunity and taking the next steps of turning it into an invention requires some attention. You may be considering putting your time, money, energy or other resources into this idea or deciding that this wasn't the one and move on to your next idea. Remember to put your emotions aside and envision when you come up with an idea that you are starting a business and this is part of the market research and development process.

Inventing Ideas

When you take on a new challenge it can take time, patience and practice before it becomes more second nature. It's similar when trying to come up with an idea or invention and may require a different approach. Be aware of your environment and pay attention to what's happening around you. Look at how people are doing things and think about how they could do it better or how you are using a daily product or doing something regularly and see how it could be improved upon.

Think Problem and Solution

How are people using their cell phones, tying their shoes, holding their purse or bag, playing a game, typing on the computer, able to see from a distance with poor vision, navigate to their destination, reducing their salt intake, making foods taste better and we can go on with so many examples of this. Ideas can be a product, service, plant, experiment, process or anything you can imagine.

Think about how the following examples evolved and look around at that the products you are using may have been modified many times to how they are being used today.

Playing football for fun without pads and having a simple catch could hurt, so Minnesota Vikings kicker Fred Cox invented and introduced the NERF football in 1972 which was a lightweight ball made out of a foam material. The Nerf Football can now be played in the dark with a light up and glow-in-the dark version Nerf football.

Look at how there is a now Yo-Yo for everyone. First, it was a toy that required a bit of skill to get that string in between the round discs which was fun for some and frustrating for others. Now there is an automatic wind up Yo-Yo that is a ball with a string inside that you throw down and returns right back to into the ball and to hand where it started from.

Did You Start on Your Invention Idea?

You think you have the next big idea, but you haven't started yet? A good way to start is to prepare. Here are some quick tips to get you started…

- Write down, text, or leave yourself a voice message describing your idea. (2 minutes)
- Draw what you think your idea would look like. Think about how it would be made, what material would be used, who would buy it, why they would buy it and what you think something like your invention would cost. (10-30 minutes)
- Do a search to see if your invention idea is already out there or there's something too similar. Check online stores, go to stores where you think your product could be sold, search websites and images online, go to the US Patent and Trademark Office website uspto.gov or google patents and search out prior patent art to start. (2-4 hours.. take your time!)

I'll check back soon! If you don't have an idea to work on… your assignment is to start noticing what's going on around you and think of simple ways to improve something you feel may be too complicated!

How Are You Doing?

So, how did it go? I'm checking in with you to see how your invention is coming along after getting you started.

Let's focus on your next step..

Did you find your big idea out there already?

Are you moving on and waiting for your next idea to hit you? It hurts for a little while and then you find your next idea to focus on without having not to have put any more time or money into it.

You didn't find anything similar and you're ready to continue moving forward? Did you take your time and really search around?

Ok good.. So now we will work on finding out if your invention idea is yours. Put it all together: links, printouts, photos of products, patent artwork you found on uspto.gov (this is the US Patent and Trademark office official site) or Google Patents website or any others.

Did you look through catalogs, maybe ones that you have laying around or on their websites?

At this point you can decide pay for a patent search.

I have a few resources that will look for prior art within the official patent databases and report their findings.

Some will give you an opinion from an intellectual property agent or attorney and let you know if, based on the prior art, it is patentable. Some search firms will just send you what they find and you have to take to an attorney or agent.

It will cost you a few bucks, but you are investing in the "research and development" side of your invention business and you want to make sure that you don't waste time or money moving forward if it turns out to be someone else's invention.

Sometimes, when you try to license it, they will do the due diligence for you, to see if you will be able to secure a patent or not. A deal breaker like that can be an expensive let-down after coming so far!

If it's not patentable, ask when the patent expires. (Design term is 14 years; Utility/Non-Provisional is 20)
If it is patentable, stay tuned…

PROTECTING YOUR IDEAS

Agreements to Protect Your Ideas

NDA's, Non-Disclosure Agreements or confidentiality agreements are typically used to put people on notice that you are revealing personal information about the idea you list within the NDA and is understood that you cannot talk about what was just shared verbally or in writing or through any exchange for a period of time without your permission.

I ask people that I am getting feedback from about my invention, people who make my drawings, engineers, Prototypers, etc. to sign my NDA. I need these people to bring my invention to a point of being able to see what it is I'm attempting to be protected through the patent process, but I do start with a patent search first.

When presenting your idea to a patent attorney or agent, you are protected by client/attorney privilege. If you want to do a double check on the attorney or agent you are considering hiring, you can go to www.uspto.gov, the official Patent and Trademark Office website and look up their registration as qualified to practice. You can also find out how many patents the attorney or agent was successful in obtaining for their clients by clicking "Search for Patents" and search by attorney name.

When presenting your invention for potential licensing, most likely you are at a point where your product has patent pending status. When I reach that stage, I don't ask the other side to exchange an NDA with me. I let them know that a search was done and that I had a provisional filed and the invention I'm about to present is patent pending.

Where to get an NDA

You can ask an attorney for a blanket template. They may charge or if you are a client or future client, they will most likely give it to you for free. You can search online for an NDA and you can also get a free copy of an NDA at www.inventorconsulting.com

One final precaution well worth the trouble...
I save all my emails, keep track of phone calls and any exchanges between receiving parties for my records in addition to any NDA's I ask to be signed.

Different Types of Patents to Protect Your Invention

Coming up with an idea and protecting it go hand in hand. You will want to talk about your invention idea to your family, friends, co-workers, potential partners or investors to share in your excitement.
During this part of the invention process you may want to discuss confidential information with others and need to have peace of mind to do so.

Below are ways to protect your idea and also work towards receiving an issued patent. If you decide to file any form of intellectual property, it is suggested to conduct a patent search first.

Many inventors may be on a limited budget or may want to explore the marketability of their ideas and have a level of protection when discussing the details. An option may be to file a provisional patent to start.

Here are definitions you should know directly from the US Patent and Trademark Office website:

Definition of a Patent

A patent is an intellectual property right granted by the Government of the United States of America to an inventor "to exclude others from making, using, offering for sale, or selling the invention throughout the United States or importing the invention into the United States" for a limited time in exchange for public disclosure of the invention when the patent is granted.

Types of Patents

The U.S. Patent and Trademark Office (USPTO) issues several different types of patent documents offering different kinds of protection and covering different types of subject matter.

Provisional Patent

A provisional patent application (PPA) is a patent application that can be used by a patent applicant to secure a filing date while avoiding the costs associated with the filing and prosecution of a non-provisional patent application.

More specifically, if a non-provisional application is filed within one year from the filing date of a PPA, the non-provisional application may claim the benefit of the filing date of the PPA.

Because a PPA is not examined, an applicant can also avoid the costs typically associated with non-provisional patent prosecution (certain attorney's fees, for example) for a year while determining whether his/her invention is commercially viable.

Further, because a PPA is not made public unless its application number is noted in a later-published application or patent, the failure by an applicant to file a non-provisional application based on his/her PPA will not lead to public disclosure of his/her invention.

Utility Patent

Issued for the invention of a new and useful process, machine, manufacture, or composition of matter, or a new and useful improvement thereof, it generally permits its owner to exclude others from making, using, or selling the invention for a period of up to twenty years from the date of patent application filing, subject to the payment of maintenance fees.

Approximately 90% of the patent documents issued by the USPTO in recent years have been utility patents, also referred to as "patents for invention".

Design Patent

Issued for a new, original, and ornamental design embodied in or applied to an article of manufacture, it permits its owner to exclude others from making, using, or selling the design for a period of fourteen years from the date of patent grant. Design patents are not subject to the payment of maintenance fees. The fourteen-year term of a design patent is subject to change.

How Much Does a Patent Cost?

Filing fees are available on the US Patent and Trademark Office website. In addition to these fees, you may hire an intellectual property attorney or agent which may charge by the hour or filing. Some inventors may choose to file on their own and need to follow strict guidelines to be accepted. See US Patent and Trademark Office Fee Schedule: [http://www.uspto.gov/learning-and-resources/fees-and-payment/uspto-fee-schedule]

Whether you decide to manufacture or license your invention, protecting your ideas could be important for you and your intellectual property. Having the security of the world to know that you are claiming your idea can give you peace of mind to move forward for the time your protection is in full force until it becomes public domain.

Once your patent is issued, it is your responsibility to protect your intellectual property from infringement and enforce your rights. Some inventors also choose to invent and receive a patent to simply hang on their wall.

Who to Trust with Your Idea or Invention

You may have this initial fear of someone stealing your great idea and that's why you've been keeping it all to yourself. The only risk here is waiting too long and then finding it in a store because someone else chose not to wait. It's up to you.

When you're ready to bring the idea to reality, you're going to have to search for outside help. This means that you will have to reveal what you are working on or thinking of at some point. So how can you protect yourself and your idea from getting ripped off? Let's see who you might need involved and how to take certain measures to keep your idea confidential with others.

Who Would I Need To Help Me?

It could be your inner circle of friends and family that you want to ask of their opinions about your idea. If you're considering filing for a patent or other intellectual property, you may want to hire a professional patent attorney or agent.

You may want to develop a computer engineered drawing (CAD) and may need the assistance of a product designer or an engineer. Perhaps you want to make a prototype and will need to send those CAD files to a 3D modeling service or get a quote for production from a factory in the United States or overseas. There may also be an opportunity to show, sell or license your invention to opportunities that come along and they may want to see everything you have been working on.

What Should You Tell Them?

At first when you are evaluating companies and what they have to offer, keep the conversation general. They will most likely need to know more specific details before providing you a quote so find something similar that already exists and mention to them to give some estimate of what you might expect. Review this process with several service providers and get a feeling of how they conduct themselves, what their costs are and what you can expect from them.

The Trust Factor

It's important to consider that before speaking to anyone about your idea or invention in detail that you have an NDA also known as a non-disclosure agreement in place. This NDA serves to protect your idea and information you share as the disclosing party and anyone who signs it agrees to keep any related information they receive from you confidential unless you permit otherwise. Whether you are seeking help from service providers, discussing your idea with consultants, or gauging consumer interest, ask everyone involved in discussions to sign an NDA and keep a record of them filed away.

You may also hear the NDA be referred to as a Confidentiality Agreement or an MNDA, which stands for a Mutual Confidentiality agreement.

Where to Get a Non-Disclosure Agreement?

In addition to your lawyer's office, there are some government sponsored organizations such as SCORE that offer a sample template on their website [score.org] (http://www.score.

org/resources/sample-confidentiality-and-non-disclosure-agreement).

You can also ask an attorney for a blanket NDA and fill in the blanks each time you present the agreement to the receiving parties. The attorney may charge a fee for providing you their form or they could charge you for each use of the form you ask them to prepare for you.

Taking the steps of researching and networking can also help with background knowledge of who you are about to collaborate with. Your risks can also be minimized by having the receiving party of your idea or trade secret on notice when asking them to sign an NDA, keeping track of correspondence and building a mutual respect for each other.

How Can You Patent Your Idea?

When you come up with a brilliant idea that you want to have commercialized, you need to move fast and patent it to avoid losing it to infringers. If you take long before you patent it, there will be two likely outcomes: one, the idea will die off never to be brought out by someone else, or two, another person will stumble upon it and make a lot of money out of it. To get the best out of the whole process, you need to know what to do to protect a novel creation.

Why Should You Know the Steps of Patent Protection?

Protecting an original idea or creation is a process with sequential steps. Many people do not know exactly how or where to start when applying for a patent.

Hundreds of applications for patents are rejected because the applicant did not know what to do and when to do it to protect his or her original creation. The first thing you need to do is find out if the idea already exists. This means you have to perform a worldwide search to find out whether it is already owned by someone else or not.

What is the Benefit of Conducting a Patent Search?

The search is usually conducted by an expert who uses advanced techniques. These techniques are not easily accessible to you. The expert has access to a database of all inventors so he or she can find out whether your application is valid or not.

Knowing what has been patented and what has not will guide you when drafting your application. Potential investors also expect you to perform a search before you submit your application.

Where Can You Get Expert Advice?

To make sure the process runs smoothly, it is best to consult with an expert in patent applications. The knowledge the expert brings on board will save you both money and time.

You will need to fill out an application form for the patent, and you may need guidance from the expert. The application procedure requires you to provide information that will guide the team reviewing your application everything they need to know. This includes abstracts, a detailed description of the patent, and anything else that will illustrate your idea and its usefulness.

How Can You Make the Protection Process Successful?

First, ensure you do a background search to avoid any conflict of ownership or duplication of an idea. You will then need to have all documentation needed to patent it. You may also need to do a little research about every step you will follow and what will be required of you.

Remember that before you gain patent status, you will have to prove the discovery's utility and its uniqueness. With all these, it is easy to have the patenting process successful.

What Are the Different Types of Patents?

You can choose a patent according to the type of invention you have come up with. Each type provides a different kind of protection for a given period.

The utility patent is the most common patent that is appropriate when you come up with a new process, machine, biological, or chemical compound.

The design patent protects the way your invention looks.

The plant patent on the other hand protects your idea if you come up with an asexual plant that has not been discovered before.

Before you go ahead and apply for a patent to protect that idea you have, you need to know some few things about the whole process. It is very important that you have knowledge of where to start and what you have at your disposal. It is also not a bad idea to consult experts who have knowledge in the field to guide you. Once you know how to proceed, then you can do so confidently knowing that you have all it takes to patent the idea.

Beware of Scams

The United States Patent and Trademark Office has previously warned stakeholders about unsolicited communications regarding maintenance fees, but it's something that bears repeating.

Many patent owners have received unsolicited communications that at first glance appear to be "official" but are not actually from the USPTO. These communications usually contain warnings about the expiration of patents for failure to pay the maintenance fees of the patent. The communications often sound urgent, in hopes that recipients will be intimidated into paying the fees listed, which frequently include the cost to maintain the patent as well as a "service charge" for the third party's trouble.

While maintenance fees must be paid three, seven, and 11 years after the patent issues, a patent owner can pay the fee without the assistance of a third party. In fact, the USPTO has made it possible for a patent owner to pay the fee online easily and securely.

Further, if you need assistance with determining when maintenance fees are due, you can check online for yourself or contact us at the Inventor Assistance Center at 1-800-786-9199 for assistance.

If you receive a letter or an email that you suspect may be deceptive, contact us via email or telephone at 571-272-8877.

You can file a complaint with the Federal Trade Commission (FTC). The FTC will not resolve individual complaints, but they may initiate investigations and prosecutions based upon widespread complaints about particular companies and business practices.

DEVELOPMENT

How Much Money Will My Invention Cost

It won't be long before you start wondering how much it's going to cost to bring your idea into the world. Investing your time and resources can add up and each invention will have its own set of requirements.

Whether you choose to manufacture or license your product, you need to know how to develop a budget and set up a financing schedule to pave the way for the expenses to come. Building anything comes with fresh sets of expenses at various stages of development. Once you have a better idea of the costs that are involved, you can plan your financing wisely.

Most people have their vision so firmly fixed in their mind that they allow their emotions to take over their business decisions. Let's review some of the steps that will cost you money so you can begin to build your budget to support your dream.

- The Patent Search: $250 to $1,500

Finding out if it's your idea can be a crucial step in moving forward. A professional patent search can run from $250 to

$1,500, depending on the people you hire and the complexity of the search. There are companies that strictly do patent searches and then provide you the information that they have found for you to review.

You may decide to take the findings and have a patent attorney or agent evaluate the patentability of your idea based on the intellectual property you gave them. You can hire a patent attorney or agent and ask them to do the patent search and provide an opinion.

Some inventors do searches themselves using Google Patents or the U.S. Patent & Trademark Office website. They go online to find similar concepts in existence to present to a patent attorney or agent for review.

Some inventors may take what they find and move forward. They run a big risk when they do this because evaluating prior claims can be complicated for those not skilled in intellectual property requirements.

If a patent search and evaluation is not done properly, there may be expensive consequences when you try to license your idea or bring it to market, only to learn that you're infringing someone else's intellectual property.

- Patent Protection: $2,000 and up

There is a less expensive option to secure a filing date for your invention by filing a Provisional Patent. You could have a patent attorney or agent help you with filing and they would charge for their time as well as the filing fee from the US Patent &

Trademark Office or some opt to do it themselves and just pay the filing fee.

Having a patent attorney prepare a patent can cost anywhere from $2,000 and up, depending on the complexity of the invention. This does not include the patent office filing fee or the lawyer's fee to review and respond to the patent office 'office actions'.

When the patent office notifies you that the patent will been allowed, you will be charged an additional 'notice of allowance' fee as well. Here is a link to the U.S. Patent & Trademark Office fee schedule.

- Engineering/Product Designers: $300 to $5,000+

What Your Invention Looks Like and How It Works
Engineering or product designers can run from $300 to $5,000 or more based on the fees they charge and the time it takes to deliver the final designs. Some may work on hourly or project based fees. More complicated concepts with electronic components and computer chips for example, can cost considerably more.

- Prototypes or Models: $50 to $5,000

Your mockup or prototype can cost as little as $50 and as much as $5,000 for a basic concept. It could be more depending on the complexity of your invention.

There are many options for prototypes, including 3D printing and model making. Producing a good works-like, feels-like prototype will allow you to bring your idea to life and test it to see how it works. It will also allow you to demonstrate to potential licensees, investors and partners can be an important step in the process.

- Tooling: $1,000 to $100,000+

Choosing to do it all yourself can mean that the greatest risks and the greatest rewards come to you, if you are the manufacturer, distributor and customer service guru all rolled into one.

Tooling costs will depend on the number and size of the component parts and materials used or needed. It could range from as little as $1,000 to hundreds of thousands of dollars. This number is determined by the product you want to produce and the country you choose to do your tooling, molding and manufacturing.

- Manufacturing/Assembly: $2,000 to $50,000+

Depending on the provider, prices can vary widely. Packaging, marketing, advertising, warehousing, distribution and staffing all come with costs as well.

- Alternatives to Manufacturing

You can choose the less expensive option of licensing, where someone else carries the costs of manufacturing and its details. When you take this route, your investment may be limited to the patent search, the patent filing costs, preparing designs and making a prototype for presentations. Your risk and return may be less but it's an alternative for those on a limited budget.

Moving Forward or Moving On With Your Idea

Did you find your idea out in the market already? Is there a big enough difference from what you were thinking to what already exists? Did you not find anything at all that looks like or describes your idea?

Claiming Your Idea

The answers to the questions above may require professional help by a patent attorney or agent for them to read through the patent records and review the products you found. You can find a list of registered intellectual property services in your area on the [U.S. Patent and Trademark Office] (http://www.uspto.gov/) website mentioned earlier.

This step may cost you a few dollars but may be well worth it. Imagine skipping over this step, starting on your idea and then finding out that you are infringing on someone else's patent. You should ask for a letter from the lawyer or agent that will give their opinion if your idea may have a chance for intellectual property protection with a patent or other form of protection.

Coming up with your idea, evaluating the opportunity and taking the next steps of turning it into an invention requires some attention. You may be considering putting your time, money, energy or other resources into this idea or deciding that this wasn't the one and move on to your next idea. Remember to put your emotions aside and envision when you come up with an idea that you are starting a business and this is part of the market research and development process.

LICENSING AND MANUFACTURING

What Does Licensing Your Invention Mean?

So you came up with a great idea and want to make some money with it but don't have too much money or don't want to start manufacturing or quit your day job… how about licensing your invention?

Licensing means finding a manufacturer with an existing distribution chain. The manufacturer with distribution is called the Licensee and the person presenting their invention is called the Licensor. You are looking to give this licensee the rights to your intellectual property or idea for them to rent or buy and bring to retail for you with little risk for you.

The Licensee can have their own manufacturing facility or outsource their manufacturing, but ultimately they control what products they are producing.

The Licensee can have a built in distribution team that presents the products to their current retail accounts or pitches new accounts for distribution. If they choose not to have an in-house sales platform, they can hire Sales Reps or Sales Agents that have established accounts in specific categories and industries.

Before you approach a manufacturer, you need to have your intellectual property protection in place. You should have done a patent search and you should have a Provisional Patent or Design Patent filed so you have Patent-Pending status.

If you want to skip all these steps and just start with a confidentiality or non-disclosure agreement, that is at your own risk and not every company wants to look at non-protected ideas.

While researching manufacturers, think about the type of stores you imagine your product in. Take a tour and try to find products similar to yours. Flip over the packages, write down the company names and get ready to start making some calls to ask if they look at outside ideas. If they say yes, ask if you could present yours. Make a prototype you can use to demonstrate how it works. Make a short video. Send photos to get their interest.

Royalties

If they are interested, they will offer you what is called a royalty, which is a percentage of sales that most likely you will receive quarterly.

There are other provisions that are included in a licensing agreement including territory, guarantees, advances, etc. Most of the time the licensee, your new partner, will involve you with certain steps of the design modifications, packaging designs, activity of production, placement into retailers, etc.

If they are not interested, try to get feedback to see what they thought of your idea and keep the door open for opportunities to present your next idea to them.

Keep looking for the right partner to license your invention to. There are licensing agents within industries that can also help you make a match. You can find them at industry trade shows and in industry magazines. You can also ask distributors and manufacturers, and network with other inventors at inventor clubs in your area, etc.

Finding a Manufacturer for Your Invention

You brought your invention through the beginning steps of the process and now exploring various options of bringing your invention to market. You could license your idea to a manufacturer with distribution which requires minimal risk and pays you a royalty or you may decide to go into business for yourself by manufacturing your product and taking a larger risk with potentially a greater return.

Responsibilities of Manufacturing

Now that you decided to manufacture your invention, there are some questions you need to ask yourself.

Do you have the finances to invest for set-up, tooling, production, packaging, liability insurance, websites and marketing initiatives?

Do you understand your industry and what the buyers or sales reps will expect of you?

Are you prepared to handle shipping, warehousing, accounting, product safety testing and other responsibilities that come along with starting a business?

There are entrepreneurs that forge ahead and find out these answers as they go along, perhaps learning lessons along the way and some may look to find a partner or partners to help in sharing the responsibilities.

Manufacturing in the U.S. or Overseas

The entrepreneurial inventor may look to keep manufacturing in the United States or search for manufacturing overseas. You will need to evaluate production costs, volume minimum order quantities (referred to as MOQ) and maximum capacities of production, shipping costs and overall comfort level of a local factory vs. overseas.

With technology today, cameras can be honed in live on your production activities whether in the US or overseas through the internet.

Finding a Local U.S. Factory

When searching for a local manufacturer, there are resources online with directories of manufacturers such as ThomasNet.com. Filter what you are searching for by using the words "contract manufacturing" and include the generic industry or specific terms that the search would recognize. Contract manufacturing means that the factories are for hire and work with outside companies.

You can also search by state and even by city to find a factory near you. Trade shows have manufacturers that exhibit and you can speak to them directly. Ask other people, including friends and family you may know that are involved in manufacturing or work in industrial parks may give you a lead.

Also, finding a fellow inventor that may have taken their invention into manufacturing may have a factory they could share that work with inventors and possibly accept smaller orders to start.

Searching For an Overseas Manufacturer

Exploring overseas factories can be just as easy as finding them locally. There are many online directories to explore such as a popular international portal called Alibaba.com.

Search for a similar product compared to yours that can direct you towards a matching factory. Pay attention to the vendors on the website while doing your search by referring directly to manufacturers. There are many distributors that simply sell product inventory from factories and there are brokers that represent many factories.

Indicate which are factories, connect with them with an online chat feature, web video call, call by phone (many of them speak English well) or email them for more information and a time to connect.

Do research away from the website on a search engine to learn more about them or go to the direct website of the factory and get familiar with what they offer.

Be sure to ask if they own the factory, if they're an employee or a broker.

Ask questions about what type of products they manufacture, get references and call US companies that they work with. Also, as mentioned earlier, check with other resources and referrals to connect you with overseas factories.

Requesting a Quote

When approaching a manufacturer for a quote, they most likely will be asking you for computer engineered drawings (CAD) or 3D files that the factory will be able to review in development software tools to evaluate how much material and labor may be needed. They may also help in assisting to identifying what type of materials you could use and estimate a cost for the tooling or set-up costs to mass produce your product.

Finding a manufacturer may be one of the most important decisions you make. Think about the time and costs associated with starting with production. Once you start with a factory, it may be complicated to start moving around your production and equipment used including tooling or molds.

You can look at your selected factory as your partner and work towards open communication, understanding and expectations to have peace of mind for you to focus on other steps of launching your product.

MARKETING

Ways to Get Your Invention Out There

Deciding how to bring your invention to the next level can lead you into different directions. You may need to pitch your soon-to-be product to retail buyers, licensees, agents or representatives and should take the time to evaluate what may be the best options for you. Do your research and make sure you are not pitching a shovel to a kitchen gadget company!

Infomercial Companies

Direct Response TV (DRTV) companies that sell products on TV are looking for mass audience appeal and will pick up where you left off in your invention process. Pay attention to infomercials you see, get a sense of how they are delivering their pitch and what type of audience they are focused on. Notice that most follow the pattern of problem-solution ideas. Ask yourself if you think your invention could be demonstrated in a way that the TV commercial delivers the message and evaluate the price points.

Before presenting to a DRTV company, do your research by visiting their websites, going to retailers that sell AS SEEN ON TV products and jotting down the companies listed on the packaging to review online.

Presenting to Your Invention's Industry
If you want to pitch an industry specific product, look for the leader brands by visiting their websites and/or call their office and find out what their process is to pitch or submit your invention. Make sure your invention idea is relevant to the company's product line.

There are agents within industries that can walk your idea to the licensees or you may want to attend upcoming industry trade shows relevant to your invention. You can go onto trade show listing websites such as [Trade Show News Network] (www.tsnn.com), search by industry, area, and region.

Once you find a tradeshow you can attend, visit the website, review sign up credentials needed and attend. Many are inventor-friendly and even have a dedicated area to showcase new inventions. When you are at the tradeshow, you have the opportunity to see new products from the industry, find potential licensees and maybe even bump into a few buyers along the way.

Catalogs and Inventions
Catalogs are looking for unique, specialty products to insert into their print and online catalogs. When you contact them and they accept your product, they purchase finished goods from you and do not license it from you.

You need to be ready with inventory to ship from or a date when your product will be ready to ship to their warehouse. Be prepared to send samples to the catalog buyers after you make contact and they are interested.

Once accepted, they will send you a vendor agreement to be filled out and get set up with them, followed by a purchase order with quantity requested and window to ship. Catalogs can be a great launching pad for your new product with smaller orders and higher margins to get you started.

TV Shopping Channels and Inventions

Home shopping channels may also be a good way to gauge interest and make money when launching your invention. Spend some time watching the format of these channels and look at their schedules for the show times that are similar to your product.

Go to the shopping channels direct websites and sift through the required product submission process or call someone that already has a relationship there that can make the process easier and quicker to get to the appropriate contacts. Finished products, demonstrations and pricing should be available to present to the buyers when called upon for next steps.

Be prepared when presenting your invention to companies. Know your product and industry, who you are presenting to and determine what your expectations are as well as your potential distribution partner.

Keep your emotions out of your decisions and put yourself in the marketer's position. Think about what's in it for them first, who their audience is and then make your decision to move forward.

Tips on Getting Your Product to Retailers

So you just manufactured a gazillion units of your invention and they are in your garage. Hopefully you had a game plan on where you were going to sell them, or you already had purchase orders to fill.

If not, here are some helpful ideas…
Look for upcoming trade shows that feature the industry your product is in and see what it takes to have a booth with your product ready to be ordered.

Look up the manufacturers' rep groups, which usually sponsor their industry trade shows and make some calls to them. These reps will look at your product, ask for samples, and take your product along with the other products they represent to buyers they have relationships with to try and get purchase orders. They get a commission and you get an order and a new account to sell your product in! You should have your product ready to go, have various packaging options ready if necessary.

Also ask these reps a few questions and do your research: what region they cover, what type of retailers they present to, what their commissions are. Remember to ask to see their agreement or discuss their terms in advance, ask what product lines they cover and what their commitment is to your product to make sure your expectations are met.

Think about talking to big box retailers, online retailers, catalogs, specialty stores, shopping channels, street fairs, etc. No matter how you sell your product, the greatest feeling is seeing your invention for sale, watching people use it and making some money from it!

Pitching Your Invention to the Pros

Are you ready to pitch your invention to the pros, buyers, licensees, agents, etc.?

You need to be prepared. Do your research and make sure you are not pitching a shovel to a kitchen gadget company!

Direct Response TV (DRTV) companies are looking for mass audience appeal and will pick up where you left off in your invention process.

I still do my due diligence and have a proper patent search done and some level of protection and I make sure they know that the product is patent pending.

Do not share your provisional patent application with them until you are in your negotiation stages and have a non-disclosure signed or a mutual understanding of some kind. I like to have it in an email that I keep on file.

If you want to pitch an industry-specific product, look for the leader brands, go on their website and/or call their office and find out what their process is to pitch or submit your invention. Again, make sure your invention idea is relevant to the company's product line. There are agents within industries that can walk your idea to the licensees.

Attend industry trade shows. You can go onto trade show listing websites such as www.tsnn.com and put in your industry and/or area/region, sign up and attend!

Catalogs are always looking for specialty products
They do not license your invention; they purchase your finished products, so be prepared to send samples to the catalog after you have made contact with one of the buyers. If you are accepted, they will send a purchase order and a vendor agreement to fill out and get set up with them. Catalogs are a great launching pad for your new product!

Home shopping channels are also a good launch pad
Uniqueness needs to be there, price and profit margins have to be competitive, and the product needs to be ready to go! Go to their websites and sift through the process or call someone that already has an in and can make it easier and quicker to get through the doors.

RESOURCES

Using Sell Sheets to Promote Your Invention

By Newsday columnist Jamie Herzlich

A well-planned and produced sell sheet can ease the way to a handshake at the end of a deal.

So you're peddling your product and you want to grab the attention of a buyer. One of the best ways to do that is through a sell sheet.

Sell sheets, if done right, can be a powerful tool to pitch your product or invention, experts say.

"It provides all the information a potential buyer would need," explains Walter Reid, a business adviser with the Small Business Development Center at Farmingdale State College, which has assisted inventors and entrepreneurs in creating sell sheets. "It initiates the ordering process."

Specifically, it's a one-page document (that could be double-sided) that describes in detail what the product is and what it does, Reid says.

It's handy to have around not only to garner the interest of potential buyers, but also as a quick reference sheet for trade shows, says inventor Brian Fried, president of GotInvention.com, a Melville consulting firm, and president of the Inventors & Entrepreneurs Clubs of Nassau and Suffolk Counties.

Regardless of how you're using it, "it needs to get right to the point," says Fried, who has used sell sheets to pitch buyers for his own inventions. Buyers have little time and short attention spans, he notes.

"You really have to catch the buyer's attention and keep it for the short time they're focused on it," he says.
Some key components of a sell sheet, says Fried, include:
— A headline to grab interest

— The name of the product or invention

— A brief description of it

— Highlights of the product's benefits

— Product images to show how it works/what it does

— Minimum quantity order

— Pricing information

— Contact information

Focus on your product's benefits, not just its features, advises Rhonda Abrams, president of Planning Shop, a Redwood City, California-based publisher of books and software for entrepreneurs and author of "Successful Business Plan: Secrets & Strategies" (Planning Shop; $44.95). Use bullet points to list key features and benefits, she advises.

"It pays to create a digital sales sheet as well as a hard copy," says Abrams, noting that each product should have its own sell sheet.

"You can send a link to your sell sheet to a buyer and add video of the product in the digital version," she says.

Providing good images is key," says Michelle Seltzer, owner of Xcessorize by Michelle in Eastport, which sells ladies accessories and receives sell sheets pitching products.
"Photographs are great," says Seltzer, who has found sell sheets helpful in choosing new products as well as when it comes time to do a reorder.

"I like to bring new products in," she notes. "It's a great approach to highlight the product."

"When sending the sell sheet, find out who the buyer is at a company and let them know it's coming," says James Shehan, graphics manager for Nevada-based inventRight, an inventors resource. "Otherwise, if it's an email, it may look like spam to them," he notes.

"Pay attention to design," advises Shehan, adding that "typically, people want to put the kitchen sink" in their sell sheet.

"Put the most pertinent information on the front and more of the specifics on the back," he suggests. "Think of the front of your sell sheet like a billboard," he says.

"Make sure the layout looks professional and there are no

misspellings," he says. It pays to consider getting expert help with it.

"It needs to look good," Shehan says.

"If you're mailing it, perhaps use a colored envelope or an odd-sized mailer," Reid says. "It should stand out a little bit and garner some attention," he notes.

5 SELL SHEET MISTAKES

1. No benefits statement, needed to show how product can solve a problem.
2. Poor design or layout.
3. Spelling errors.
4. Trying to cram in too many details.
5. Shoddy product images.

Inventor groups can really help you

There are inventor groups everywhere and you may not even know it. What a great way to meet like-minded people with ideas for new businesses and inventions.

When you attend the usual monthly meetings, you can find local resources, share referrals of patent attorneys, Prototypers, designers, licensing agents, etc., network, hear from speakers and I'm sure everyone takes something else from them.

I run the Inventors and Entrepreneurs Clubs of Nassau and Suffolk Counties for both County Executives (meetup.com/ienassau and meetup.com/iesuffolk) and it is amazing to see everyone continue to move their ideas and inventions forward.

It is great to hear of the success stories and the journeys inventors take to get there. It's not the easiest process, and everyone can have a different path, but ultimately we all have a similar goal: seeing others use our inventions and making money from it!

I advise you to find an inventors group in your area immediately and attend one to give it a try. Here is a head start on your search:
uspto.gov –and click on inventor resources area

If you don't have a group in your area, contact me and I'll

refer you to a great interview on Got Invention Radio (www.gotinvention.com)- lookup the interview with Terry Whipple, who has a DVD on getting one started.

If you do have a group in your area, show your support and show them that it can be done...become the next success story that comes out of your group!

Finding an Inventor Mentor

When you come up with that big idea there may be questions you need help with or during your invention journey, there may be times where you may feel like you just hit a brick wall. You scour the internet looking for answers from the right person or company to help you and are hesitant to get anyone else involved. At some point on your venture, you will need to connect with someone to help you take your invention to the next level. A mentor can be there to coach you through the invention process, help make decisions and connect you to various resources.

Characteristics of a Mentor

A mentor is someone who has walked a mile in your shoes. Because he or she has stood where you are standing now, the wisdom you can acquire from a mentor is not random knowledge - it is specific to your needs. A mentor can be a sounding board that you can share your ideas in their formative stages and gain valuable insights and fresh perspectives from someone who knows where you are coming from and where you hope to go. A mentor holds the key to your future networks of colleagues and collaborators. His or her stature in the community will leverage your credibility; introductions he or she makes will open new doors for you. You can look to them as a friend, someone who has your best interests at heart; someone whose agenda includes you.

Where to Find a Mentor

You will come across likely candidates each time you speak

to an established professional whose work and work ethic impresses you. How do you get them to help you? You ask. You'll be surprised at how eager many of them will be to invest their time and energy in someone who is working hard to achieve a dream. Why will they do that? Because someone once invested their time and energy in them.

Another way to find the mentor who may be helpful to you is to think of the people who have the most impact on you when you're all alone, reading and researching. I'm talking about the authors whose books, articles or opinions make the most sense to you. If their words have that kind of influence on you, imagine how helpful it would be if you could pick up the phone and get their insights and advice when a particular issue completely baffles you!

Connecting with a Mentor

In many cases, mentors are in business themselves. Their contact information is readily available, either because it is attached to their book, their article or their blog. You can also contact the source that published them. Newspapers and periodicals, information websites, TV and radio shows all have public access information. Publishers usually have an "About the Author" page on their websites as well.

Working with a Mentor

Before you pick up the phone or send that email inquiry, however, make sure you know what you'd like this individual to do for you. Many experts offer their services for an hourly fee, so don't be surprised if your query generates a price list of consulting services in return. Others may have agendas and are using their books and advice columns to build ancillary businesses.

This may not be the kind of relationship you are looking for. Still others may be willing to work with you in exchange for a percentage of the royalties your product will earn; they may even be interested in a percentage of your business, if you have more than one concept in development and your company is well grounded.

You are going to be connecting with design engineering, licensing, manufacturing and distribution while working on your invention. Every now and then you will be speaking to someone you feel a real connection with and you'll hear that little voice inside your head saying, "I wish this was my mentor!"

When that happens, the first thing you should do is ask them if they ever take on the role of mentor and if so, would they be interested in mentoring you. In exchange for their help, you can offer them percentages on your royalties and/or in your business.

INTELLECTUAL PROPERTY EVALUATION TOOL FROM US PATENT & TRADEMARK OFFICE

While looking for great resources to share for inventors and entrepreneurs, I came across a free online tool from the US Patent & Trademark Office, that will allow creators of intellectual property to recognize when they have an asset that can give them a competitive edge in the marketplace and when they should seek IP protection. Find out more about various patents you can file for, trademark or service mark protection, copyrights, trade secrets, licensing technology to others, international rights and other assessments of potential protection.

Check it out...this IP Awareness Assessment Tool enables you to measure and increase awareness of IP issues, relevant to creative projects and your business goals. Take a few minutes to answer a comprehensive set of questions regarding IP (Intellectual property), after which the tool provides a set of training resources tailored to your specifically identified needs. The tool is available on USPTO's website atwww.uspto.gov/ inventors/assessment/.

Website Resources

USPTO.gov
The US Patent and Trademark Office website is a must for you to check out and take a peek at the inventor help tab. Search patents and trademarks, find prior art and share it with a patent attorney or agent for an opinion.

Score.org
This is a group of retired executives offering their help for free within industries

SBA.gov
The Small business Administration website with resources for businesses

InventorsDigest.com
The only magazine for inventors

UIAUSA.org- great resources

GotInvention.com
This is my live weekly online radio show for inventors with over 1500 archived interviews for you to download or listen to anytime. I started the show in 2009!

INVENTOR INTERACTION

Q&A's with Brian Fried

What are 3 trends from inventors that excite you?

1. I see more and more companies looking for new, fresh and exciting ideas from inventors.

2. I see more and more people looking to pursue their ideas because of today's economy, and they want to set themselves up to have residual income or start their new business.

3. I see more and more people needing great information to help them with their ideas, and that excites me to host the radio show, Got Invention Radio.

How do you bring ideas to life?

I quickly write my idea down to capture my thought. Then I do a quick search on the major search engines to see if it already exists. I make a great prototype, get some patent protection and explore the options of manufacturing or licensing. I also keep exploring my options and getting the word out by going to trade shows related to my product and connecting with media resources. I keep exploring options until I make that connection that will bring my product to market.

What is one mistake you've made that other inventors can learn from?

I have many mistakes that I learned from, but I have to say one major one is that you have to talk about your idea and connect with resources to get somewhere with your idea. Too many inventors don't want to talk to anyone about their ideas. That was me, but not anymore. Do a search, get some intellectual property protection and start talking about it!

What is one book and one tool that helps inventors bring ideas to life?

Can I mention my book? "You & Your Big Ideas" was written by me in order to help inventors. It's all about you and your big ideas! I wrote it so people can have a resource guide that takes them through a step-by-step system, while giving them great tips and resources to save them time and money. And a resource tool for all to use is the archived audio files available for free of past interviews on Got Invention Radio. There are hour long interviews for every aspect of the invention process.

What is one great inventing tip that you're willing to share with inventors?

Go to the store and then to the section that you see your product being sold at. Then look at the products that are there. Write down the company names that have a similar product to yours, and then go home and do research on that company, and call them to ask if they are interested in looking at your idea. You can find a partner that already has manufacturing and distribution, otherwise known as a licensee, and get a small percentage of a bigger opportunity, with little or no risk financially.

What's next for you?

I would love to host a TV show for inventors. I was asked by inventors to work on a solid show, just like Got Invention Radio, interviewing top experts and resources. I also plan on taking my inventor consulting business to the next level, providing inventors with solid help and advice from scratch on a napkin to seeing their invention ideas on TV, catalogs, mass retail, home shopping channels and online retailers.

How do you feel about inventing?

It definitely has its ups and downs both emotionally and financially, but it's a passion of mine to find solutions to things that I can see being done differently or made easier. It's just the most amazing experience to see your product out there for someone to buy and use it for the same reason that I came up with it.

What should an inventor keep in mind when starting out?

- Track the time and resources you're investing in your product, and keep them separate from personal finances.
- Consult with a patent attorney or professional search firm to make sure your idea can be patented, and to keep the process on the right track.
- Be alert to your costs and market forces, and how they may change in the future.
- Present a clear, concise pitch to potential partners that shows the unique value your invention.

What happens if you find a similar idea has already been patented?

Your invention may be novel and unique enough that you can still move forward. You may also consider working with the company that holds the patent. But don't do anything until first consulting with a patent attorney. Another option is come up with a good "trademarkable" name and start manufacturing, but make sure you are not infringing on someone else's patent!

What are some good ways to monitor the pulse of the market?

•Research constantly. Trade shows are a great way to learn about and gauge up-and-coming trends. Industry trade publications and websites will also help you stay involved.

•See what competitors are doing. Visit retail or distribution channels where you feel your products should be, and talk with their owners. That will help you stay in the loop.

Quotes and Inventing Inspiration

"All achievements, all earned riches, have their beginning in an idea." Napoleon Hill

And this one was from my fortune cookie!
"Happiness lies in the joy of achievement and the thrill of creative effort"

Keep moving forward with your inventions. You are almost there!

DailyOM
Inspirational thoughts for a happy, healthy, fulfilling day

"Each time a visionary shares their ideas, there may be some struggle though there is also likely to be evolution. Our goal here on earth is to become our best selves and to help others do the same.

"Inspiration is the universe whispering to us. We act as its hands and feet, bringing it from the realm of thought into manifestation. Part of our journey here is to experience the universe by working with concepts and energy on the material plane."

"You could discover innovative ideas today, which could leave you feeling resourceful. It may be that you have noticed that things can be done much more efficiently, and you may feel that you have hit upon the perfect solution.

In order to develop your inventiveness, you might work on looking outside the box. As you work, you can look around you and notice any problems you see. Then you may want to think about your immediate reaction to the problem and write it down.

This will probably be something that has been tried and true. Next, you can imagine that you have encountered this issue from a different perspective and brainstorm a few solutions that seem completely out of the ordinary.

Combine all of your ideas to come up with a new solution. You may find that giving yourself time to think in this way helps you reinvent the wheel.

Allowing ourselves to engage in other ways of thinking gives us the means to find creative solutions to a problem. Our usual methods of troubleshooting can limit us, because we tend to get into a groove and stay there.
However, when we picture ourselves or our problems from an outside perspective, we discover that there is more than one way to generate solutions. Our ideas grow the moment we give ourselves permission to see differently.

By changing your vantage point today, you will see that if you expand your view you will come up with original and inventive ideas that will make things work more smoothly.

How do I become an inventor, Mr. Brian Fried?

I was thrown off by the question for a moment and it kind of made me go way back before even coming up with the idea! So I answered like this…

Look around you! You need to be aware of your environment and when you enter different environments and just pay attention to what you are doing that could be totally new or modified to a point where you can claim that it's your idea.

When you are placed somewhere out of your comfort zone, see what people are doing and how they are doing.

Find better ways, find an improvement, make anything easier, come up with something for the first time...

Imagine it in your mind and make it real!

Come up with a name for it to have an identity.

Get online and do a search in online retailers and search through search engine websites with descriptive variations of what you just thought of and see if it's easy to find. Take your time and see what comes up or similar to it.

Then go to uspto.gov and/or google patents and do the same-search out existing patents and prior artwork of those inventions that have or have had patent protection in the past.

Keep a log of what you find and then I would suggest taking it to an attorney for an opinion if you're on your way to owning a patent!

**Once you start, you'll discover that there is a whole new way of looking at everything.

MONEY SAVED-LESSON LEARNED

Recently I was consulting an inventor on Skype that was just starting off with her invention. She had a homemade prototype, quote from a patent attorney of $5,000, a quote from a patent searcher of $1,000 and a quote for engineering of $9,000.

She struggled to explain what the purpose of her invention was at first and once I understood it's purpose and function, I put up a red flag. Out of all the inventions in the world, she presented one so similar to one I've already worked on with another inventor that was fully patented and in retail already. I showed it to her on a website and she agreed that it was just about the same thing. Was that a gift?

We discussed what her options were which was to work with the existing patent holder or modify it so it would be unique in its function, but most likely would be a far road taking that direction to make improvements on the original invention and calling it her own. Eventually she was coming to terms and telling herself that she was hitting a brick wall.

It hurts at first and for a while after, but I told her to take a look at those quotes from the patent attorney and engineer and think about all the time and effort she has and would have continued to put in to hit that brick wall- money saved, lesson learned.

Whether it's going to the store, online, doing your own patent search, finding out from a professional search, finding out if it's out there to start, will help you either find a better way to make it work(utility patent) or look (design patent), as long as you're not infringing on someone else's intellectual property. You can also find out if it's in the public domain(if the patent expired) and come up with a great trademarkable name and start to manufacture it on your own OR....move on to your next idea!

Once you come up with that first idea, the next one just happens and the next.. Clear your path for your next idea which could be the winner!

A Question on First Steps

I just love getting emails from people with their big ideas! No matter what it is, in my opinion, you need to start with doing some basic searching. Check the places you can visualize your product being sold in. Online, catalogs, go to the big box stores, specialty stores, etc. and check to see your product is already out there or a similar product exists.

Then, I would say to go online, so a solid search engine search on several sites, such as Google, Yahoo, Bing, etc. and put in some descriptive terms of your idea...see what comes up and make a note of that website address.

Also, go to uspto.gov and put in variations of the description of your idea and see what hits you get.

If it's so similar, you need to find out if the product is still under patent protection, if it's expired by being public after the 20 year patent, then you are free to work it under a good trademarkable name.

If it's far off from the claims, you may be able to get patent protection. If it's so close and protected, move on to your next idea.

Next, I would take all that you located and pay for a patent attorney or agents time to review what you found and let them compare your idea to your search results and get an opinion from them on whether your product is patentable and you have a green light to move forward, if you are infringing, or it's public domain and you can market it with that great name you came up with..

Companies you present to potentially license your idea will most likely want some patent protection, otherwise what do they need you for. They can do it on their own!
Otherwise, consider going into business for yourself.

The choice is yours!

Protecting Yourself

Today's lesson came out of a discussion I had with an inventor.. There are people that want to help inventors and take care of the entire process. You may see these ads on TV or run into people or speak to companies that will just "run" with your idea... And your wallet!

3 lessons here...
1. I would stick to specialists in each area.

I would use a professional patent searcher or bring my patent search to a patent agent or attorney and file my IP (intellectual property, including provisional, non-provisional or utility, even design patent). I would go to a professional to draw my product, hire the right sales reps, etc. I want to go to a specialist for each step and have them do what they are great at and then move on.

Many of these places that you call tell you they will get you from A through Z and act as a project manager. They have a right to get paid, but once you understand that you can really do these steps on your own, you can use that money you spent later on in the process.

2. I think you've heard me say or write this before, but I'll say it again: Do your research!

No matter what you choose, you need to ask questions. You

may have revealed your idea to someone, but don't worry, it's your choice to work with them or not.

You feel that emotional attachment to your invention, and you have to separate you emotion and be a business person when you are going to people asking for help. You need to do your due diligence and be sure you have the right people working with you on your team.

3. Ask questions.
How many patents have you issued?

How many products have you brought to market, Mr. or Ms. "I'll do everything" and where I can buy those products right now?

Can you provide me 3 references of inventors you have worked with?

You are in control–go in with this mindset and it will help you make better decisions!

A Question about Video Marketing

To start with we discussed making a video (as easy as using your cell phone) of her prototype to focus on the main points of the product that really should come down to presenting the problem(s) and your new innovative solution. Talk about all its great features, what it is or could be made out of. It should be about a minute to a minute and a half and that's all.

This will have you prepared for presenting an as seen on TV companies, home shopping channels and possible licensees within the industry your invention falls into.
Then upload the video to YouTube as "unlisted" so only people you send the link to will be able to see it. If you leave it "public", all could see it and then there is one more option where you can select "private" where only people you invite can see it.

This has been success for me many times and may be for you!

Have fun making the video and save the bloopers for future entertainment!

An inventor's question on Launching

An inventor called to talk about her product. It's fully developed and she's at a crossroad on what to do...

She has a full time job, has limited cash and wants to manufacture her product on her own. We discussed her options and spoke about starting slow, maybe reaching out to catalogs, building a website, generating online sales and building a track record and at the same time. This is a way for you to gauge interest without quitting your day job and putting your bread and butter at risk.

We also spoke about how getting into retailers is difficult, but not impossible with just one product to pitch them. The other option would be to try to license it and collect a royalty with little risk to you.

Licensing is another option. You contact a company that is in the same product category as your invention and if they decide they are interested, you can agree to terms of an advance, royalties, territories and time.

Is Your Idea Yours?

Coincidentally, I received 3 calls from inventors yesterday all at the starting point of their inventions. They came up with an idea and didn't know what to do next.

I'll give a quick answer here just for us to get past this step... Go online, and go to 3 search engines, Google, Yahoo and Bing and put in a description of your idea in several ways. If it's a pen with a light on top, look up "pen with light on top", "pen with LED" "light on top of pen." I'm just giving a raw example here and you must take your time! Spend some time going through the search engines, maybe check out some of the online retailer's stores to see what similar types of products they carry and if it's the same as yours or very similar.

Also, go to USPTO.gov to do the same. There you will find similar patents and pay attention to the art, then look at the links to the referenced patents that are listed and check those out. That will keep you busy for a while as the references and links to other patents will keep going. See what you find!

Make a list of websites, products, patent numbers, etc. and bring your findings to a professional- a patent attorney or agent and let them give you an opinion if your idea is patentable or not. Remember.. It's only their opinion, so you might want to get a second one as well.

Ask them to not only look through the artwork from the patents you found, but also look at the claims and compare it to what you came up with. You can also have a patent search company, patent agent or Intellectual Property attorney do the entire search process for you.

Take these first steps and then continue your journey... you're almost there…

How You Can Succeed as an Inventor

Quick Inventing Tips from score.org

What should an inventor keep in mind when starting out?

- Track the time and resources you're investing in your product, and keep them separate from personal finances.
- Consult with a patent attorney or professional search firm to make sure your idea can be patented, and to keep the process on the right track.
- Be alert to your costs and market forces, and how they may change in the future.
- Present a clear, concise pitch that shows the unique value your invention. What happens if you find a similar idea has already been patented? Your invention may be novel and unique enough that you can still move forward. You may also consider working with the company that holds the patent. But don't do anything until first consulting with a patent attorney.

What are some good ways to monitor the pulse of the market?

- Research constantly. Trade shows are a great way to learn about and gauge up-and-coming trends. Industry trade publications and websites will also help you stay involved.
- See what competitors are doing. Visit retail or distribution channels where you feel your products should be, and talk with their owners. That will help you stay in the loop.

MORE FROM BRIAN

How do I feel about inventing?

I get asked this question all the time.

It definitely has its ups and downs, both emotionally and financially, but it's a passion of mine to find solutions to things that I can see being done differently or made easier.

For me, it's the most amazing experience to see my product out there for someone to buy and use it for the same reason that I came up with it.

Something else that would never have happened if I hadn't become an inventor is getting to meet all the incredible people who are passionate about inventing, just like me.

For the last 7 years, I have had the opportunity to speak about my journey through the invention process to groups at public libraries and inventors clubs. I've been involved in programs from the Small Business Administration and Small Business Development Center and I've been invited to speak at seminars for many wanna-be and up-and-coming inventors of all levels and ages around the country. The seminar is called "An Inventor's Adventure" and in it I share my successes, challenges and adventures that are all part of becoming a seasoned inventor.

Since my adventure began, I've had patents issued and

products licensed. I've had several of my inventions manufactured and I've been invited to be an on air guest on QVC to promote the sale of some of them. It's a proud moment when someone on National Television introduces you as a successful inventor. I still have to pinch myself when I realize they're talking about me.

Because the inventor's life can be a struggle, and I've had my fair share of disappointments and failures, I know how it can feel when there's no one to talk to who really understands. That's the main reason I started writing and speaking on the subject, and why I worked so hard to get support for the two Inventors Clubs I founded in Suffolk and Nassau Counties.

In the midst of raising a family and taking care of a full-time job, I invested almost a year to write my first book, *You & Your Big Ideas* (youandyourbigideas.com) because I kept meeting so many people who had no idea what to do or how to follow through on their big ideas. It was very rewarding to get their emails thanking me for saving them from making costly mistakes and for being so generous with the kind of information they'd never have found without my help.

I always tell them the same thing: When it's your turn to share what you know to someone just starting out, remember how good it felt to have someone with experience in your corner and take the time to pay it forward, just like I did.

My daughter, Alana, has grown up watching this crazy inventor's life up close. She's curious and when she asks lots of questions... why doesn't this work as well as it could and why didn't someone think of doing that another way… that's when I realize she's acting like an inventor and it makes me smile.

I'm always telling her, if something annoys you, see if you can figure out another way to make it work, and then write it down. Describe the new way you think it can work. Draw a picture of what it looks like from the image in your mind. "Let's do a search on the internet and see if it's out there," I tell her.

If we find it online or in a store, I buy it for her, so she can see that something she imagined could be made into something real.

If we don't find her idea anywhere and it has potential, we start the invention process together.

Thanks to the coverage I've been getting in the local media, partly a result of my weekly radio broadcasts on Got Invention Radio (gotinvention.com) and partly because of the products I've invented that are being sold on TV and in stores, I've been invited to speak in elementary, middle and high schools and at colleges, career days, entrepreneur clubs and science classes.

Sometimes the local media covers my visits and this means a lot, because the attention is a kind of pat on the back – an encouragement to keep on inventing.

From the feedback I've been getting, this attention is extremely rewarding for the students and teachers who are opening the door and allowing the seeds to be planted for innovation in schools.

The thing is, if I can do this, you can do it too. You can share your message in your community, too. Your experiences – good and bad – have huge value to people who just need a little help now and then; a pat on the back, maybe; some advice you can offer that will inspire them to keep their eyes and their minds open for change.

Believe me, you really do have the power to excite people about possibility. All you have to do is speak from the heart.

If you're thinking about giving a talk, here are a few tips that may help.

No matter whether I'm talking to elementary school kids or college students, I've found that everyone is curious about inventors. So that's where I start. I tell them I'm an inventor and then I ask if they know what that means and what an inventor does.

The next thing I do is show them some of the things I've invented. I usually start by showing my Balloon-O-Band (balloonoband.com) because it's such a simple idea and everyone gets it right away.

The Balloon-O-Band is a nylon wristband with a metal ring on the outside where the balloon ribbon is attached. It has Velcro ends to make it easy to adjust the size and to put it on and take it off with no fuss.

One of the problems it solves happens almost every time a child wants a balloon, at a party or an amusement park: the balloon is lost when the ribbon comes loose and a happy day becomes a day full of tears.

Another problem it takes care of is the constant struggle the parent goes through, to tie and untie the balloon when the child wants to go on a ride or use a bathroom or get something to eat. As soon as the balloon is tied to the band, the band solves everything.

After the Balloon-O-Band, they always want to see something else. I usually show them the Pull Ties at that point (pullties.com) because this is another invention that solves a simple everyday problem.

When I finally got fed up with all the money wasted on stale chips and crackers and those knots I could never untie from our plastic grocery bags, I came up with Pull Ties. I made them dishwasher and freezer safe and easy to use, so no one had to struggle.

By this time, everyone gets it. An inventor is someone who looks at ordinary, everyday things that don't work as well as they could and figures out a way to make them work better.

Then I tell them that everything we see and use was someone's idea at one time. We talk about how and why people came up with some of the things people use.

Then I bring out a yo-yo and ask someone to show us how it works. When they have to rewind it by hand, I tell everyone to pay attention to what the person has to do to wind it back up and see how long it takes.

Then I take out an automatic rewinding yo-yo and I tell the class that someone got tired of rewinding their yo-yo and came up with this solution. I ask if anyone wants to try it. After that, I bring out a yo-yo that has lights and glows in the dark.

By this time, they totally get it. They realize that if something as simple as a yo-yo can have so many variations and improvements made to it, there are all kinds of inventions to improve other things just waiting to be thought of by someone just like them.

I show them a regular football and foam football and I throw it to the kids and have them toss it around to see the difference between a football that can hurt when you throw it and one that is soft and safe to play with.

Then I bring out a couple of products that have not made it to retail yet. I tell them that maybe the product could use some improvement. This is what it does, I tell them. Then I ask, what do you think could be done differently? What would you change and what would you improve?

When I go around the room, it's amazing what they come up with. Then I tell them, "Do you realize that all of you are inventors? You've looked at something and you've come up with new and better ways it could work. That's all it takes."

Then I get them excited. "Think about this: If you can't find something that matches an idea you have in your head, you could be onto something really big!"

After that, I bring a book in called "The Kid Who Invented the Popsicle" which has quick stories about how and why inventions such as gum, ice cream cones, the zipper and many more came about.

I encourage them to keep asking questions and thinking about new ideas and sharing them with their parents.
Then we do a question-and-answer session.

At the other school levels, I like to talk about being aware of new ideas that are always turning up in the stores: new technology, web ideas, new businesses, etc. and what they can do when they come up with an idea.

I walk them through the process of doing a web search, explain what a patent is, how to make a prototype and the different ways to bring an idea to market.

I explain what licensing is – that it's an agreement you make with a company that already has manufacturing capabilities and retail distribution – and I talk about manufacturing on your own, and the risks and rewards.
Estimated time: 30 minutes (can be adjusted based on the class time availability).

When you're ready to get started as a speaker who, here are some helpful suggestions:

If you have a child or children in school, ask if they have time for a brief presentation on inventions, innovations or entrepreneurship during the school year. If you know anyone that has a child in school, find out if they would be willing to introduce you to the teacher.

You can also offer to volunteer your story experience and advice to the school officials and let them know what you have to share with the students. Make the same offer to your public libraries, community groups and local government offices.

Remember: You have the gift of being an inventor, so share your gift. Plant the seeds of innovation with the next generation.

Are you someone with "a million" ideas?

It seems like I've been hearing people say this a million times. What they all have in common is that they're not sure what to do with them, or even where to start.

Let's begin by focusing on all these ideas you have. That's how I started. My brain was constantly seeing so many things I wanted to improve, modify, wanting to create the unknown, until there was a point that told me to STOP and focus.

So I took a piece of paper and wrote down all my ideas. I couldn't believe how many I had that were stored in my brain from all those years! When I started to sort them out, I realized that they were so wide ranging, the number of categories ran off the page.

After that, I put them in a simple spreadsheet on the computer and that gave me a chance to prioritize them, add notes to them and put an action plan on each one. After that, I got to do one of my favorite things – give each one of my inventions a name.

As I was going through the process of doing patent searches, patent applications, prototyping, marketing initiatives and keeping tabs of who I spoke with, which licensees and/or buyers were interested and who I made deals with, I kept referring back to my spreadsheet. This way, I was able to keep everything at my fingertips, with easy access any time.

So, I guess the bottom line is pretty simple.

Organize, prioritize and take action with your ideas.

Ask others their opinions and think about which ones have the best window of opportunities!

Get started and imagine seeing others using your invention-turned into product for the same reason you came up with them in the first place!

Trends that Excite Me

1. I see and hear of more and more companies looking for new, fresh and exciting ideas from inventors.

2. I see more and more people looking to pursue their ideas because of today's economy, and they want to set themselves up to have residual income or start their new business.

3. I see more and more people needing great information to help them with their ideas, and that mean I get to help them every time I host my radio show, Got Invention Radio, and offer my tips at inventingtips.com .

ASK THE EXPERTS

When I launched my weekly radio show, Got Invention Radio, my goal was to create a connection between inventors and invention experts.

By setting up interviews with people who make it their business to facilitate the invention process, and making it possible for listeners to tune in live or stop by the website when it suited their busy schedules, I felt it was a big step in the right direction.

With my first book, and now, with this second book, I've reached out to my best resources once more. With their permission, I've included some articles they've written that can help you make sense of this crazy career path you've chosen to either make a living or make life easier to live.

How to Come Up With a Profitable Invention

Identifying a Problem

When you decide you are going to come up with a brilliant invention idea, you are planning an invention; this is different from having an epiphany one day. Well, if you are interested in inventing, then you have come to the right place, here is an article to get you started and generally guide you through the invention process. The first step in coming up with a profitable invention idea is identifying a problem in society.

Finding a Problem

First, you have to find a problem. What is a problem? I think most everyone knows what a problem consists of, especially a problem when we are on the subject of inventing.

A problem is anything that hinders one from reaching a desired goal or object. There are problems all throughout our society, as you know. You probably encounter around 20 problems a day and most likely openly complain about half of them.

Up until now you haven't noticed when you did this, but going through day to day life as a human is your number one supplier for good problems. Listen to yourself, when you complain think about if it could be fixed with a nice invention.

It's hard to be focused enough to notice when you complain or see a problem; however, with some practice, it starts to become more noticeable. Also, I recommend carrying around a little journal or using a phone to document these ideas when they come to you.

When looking to identify a problem there is also another resource that is readily available – people around you. This might include your co-workers, friends, family, etc. People naturally complain, you hear them every day, and up until now you probably thought it was a bunch of annoying nonsense.

I've developed my own explanation for why people complain. They do it to make advancements. When enough people locate problems and vocalize the need for the problem to be solved, someone finds a solution and the next thing we know, we end up with an invention idea or advance in technology.
So, listen to the people around you, they will tell you problems they have in their lives. This can be even harder to do then listening to yourself, because we have been conditioned to not pay attention to people complaining.

The internet is a great source for information, use it. People have a problem; they post it on the internet. This is similar to listening to people around you, it's just people that are farther away. There are thousands of blogs and forums where people have jumped on the internet and posted a problem they are having. Go on Google and search for household problems or something along those lines and you will surely find something.

Also, as I talk about later, a great problem to identify is one that causes death; therefore, it could be beneficial to search online for things that are causing a death toll every year. If you start to master using these three sources of information for identifying problems, then you will soon have too many problems to remember.

Is the Problem Common?

Just so there is no confusion, you do want the problem you identify to be common. You don't want to be the only person having that problem, or else the invention idea you come up with to solve that problem will only be useful to you.

There are easy ways to decide if a problem is common:

1. Ask people you know. Talk to friends, co-workers, family, anyone you see on a regular basis. Ask them if they have the same problem. You don't have to tell them you're thinking about coming up with an invention idea to solve it. Just say, "Man, I hate it when (blank) happens. You ever have that problem?"

2. Again, use the internet! If the problem is common, then lots of people have posted about it on the internet. Do a search on Google and see if the problem turns out to be very popular.

3. Last, you can do a survey. Go somewhere that lots of people gather – a school is good; a community center is good; a team event is another good one – and get permission to ask them to raise their hand if they have this problem. This can be a little more intimidating to some people, and it will definitely let your secret out, but it's a great way to get fully submerged in your project.
Don't take this step lightly; it is very important that your problem is common. A profitable invention idea has to appeal to a large amount of people.

Has the Problem Already Been Solved?
Obviously, this is also a very crucial thing to recognize. Solving a problem is going to do you no good if there is already something that solves it. So, you have to do a little research to verify that your problem is waiting for you to solve.

As you would probably guess, a great place to start is the internet. At this point you've probably already searched for your problem on the internet, so if it has been solved, you should have noticed. I would still do a second detailed search to make sure you didn't miss anything the first time.

Next, it would be wise to do a patent search. You can do a patent search online at the USPTO's website or with a patent attorney. For now, I would recommend doing the search yourself as it will be cheaper and easier.

Searching for a problem rather than an invention is pretty difficult, so it may take a while. Also, you can try searching an obvious invention idea that applies to that problem for better results.

I find it helps a lot to ask around. A lot of times someone will say something like, "I think I heard about something that does that," or something along those lines. Don't forget the value of people.

Is Solving the Problem Viable?
The ultimate question: is the problem going to be profitable to solve? There is no definite answer to this question. There are many things to consider once you have made it to this question:

Does it save people money?
This is a big one. People love products that pay for themselves. We are also in the midst of an economic recession, so people are always looking for ways to save money.

If your goal is to save people money, your invention has to be very cost efficient. People don't want to hear that your product will pay for itself in five years. They want quick results.

Does it save people time?
The average person hates spending time on things they don't enjoy. People want to get back to their free time, so saving them time can be very appealing. Over the years, time becomes important to us for many reasons.

Does it create comfort or entertainment?
This is a difficult question to answer, because it is very subjective. People get comfort or entertainment from so many different things. So if you choose this qualifier, you have to be sure your invention will apply to lots of people. If you are going to attempt to tackle this, I would recommend doing extra research, online and in surveys.

Does it save lives?
This is another big one, and it can lead to very profitable invention ideas. There will always be a great way to market an invention that saves lives. I would definitely recommend going down this path.

Does it help people who are handicapped in some way? This is a broad category and almost goes with saving lives. There are many things that handicap people, such as allergies, diseases, physical limitations, etc.

Would people pay for this problem to be solved?
To really answer this, take some thought. You have to consider many variables. You have to think about how much an invention to solve this problem would cost and if that amount of money is worth solving the problem.

You have to consider your consumers. In today's economy people are spending a lot less money, so you need to do some research to find out what people are still spending money on. The worst thing is to have a great invention, but not be able to make money due to a lack of research at this step.

Now you should have a favorable problem to solve; you are on your way to a profitable invention idea.

Problems Create Ideas

Many of us look at problems as a negative thought process and others look at those challenges as opportunity. These problems can be the start of solutions, finding ways to make what that problem was easier and better.

Start to write down what you think the solutions can be. Is what you are thinking of make it easier to accomplish? How and why is your solution better? You are most likely asking and answering these questions to yourself. Think about checking-in with the public from a broad perspective and see what they think.

Continue to Keep Your Solutions to Yourself
You want to do some general research and figure out if your solutions will be something that others would potentially use. Go online and lookup what it is you are dissatisfied with and keep your solution in mind to evaluate if others are complaining about the same thing.

Are they talking about why it's happening or what should be improved? Ask people if your problem is happening to them. Keep a list of your research and findings and evaluate if this idea has potential.

Captivating Ideas

Ideas can come and go. As I've said before, you need to make a record of them. Then, you should organize them into categories such as housewares, hardware, apps, novelty, scientific, electronics, toys, games or other categories related to your ideas. After that, prioritize your ideas based on the ones that have the most potential and cost the least money to develop.

Practice this exercise of focusing on problems and solutions to be more aware of the things people want solutions for and you will soon be taking your idea and turning it into an invention.

Uncovering an Invention Idea

Use Problem Solving

All problems require problem solving; however, some are easier to solve than others. An invention requires a high level of problem solving; this is why it is hard for most people to come up with a unique invention. Some people are naturally good at problem solving and they can solve problems with ease, but there are many people who have developed ways to teach people how to solve problems.

There are many great sites and books that give full lessons in problem solving and I would recommend checking some out. I am going to outline a couple of things that I think are important to problem solving in inventing.

Completely simplify the problem.

It is easy to look at the problem and get caught up in the complexity of solving it, and this can discourage you. Take the problem you have identified and find the underlying cause. You want to have a simple problem in order to have the simplest solution. Simple inventions are generally very profitable.
Change the setting of your problem.

This might be difficult to understand, so I will give you an example. Let's say your problem is getting in the shower and then realizing you forgot to get the new shampoo bottle and now you are all wet. Okay, now change the setting from a shower to your car. The problem here is that you forgot to get a new pack of gum to put in your mouth before you leave to go on a hot date. Doing this will allow you to look at the problem from multiple angles. Don't be afraid to do this several times for several different settings.

Problem solving is about solving a problem by using logic and using techniques to enhance your logic. However, logical problem solving isn't the only way to go about solving your problem; you should also consider the abstract side of pinpointing ideas, which I will talk about next.

Imagination

Imagination is built almost completely around vision. Vision contains two elements, one of which is often overlooked. The first part of vision is the retina receiving light rays and sending them to the brain, this is this part of vision that everyone is familiar with. The second part, what happens after this, is the part that relates the most to imagination. The part of vision where the brain interprets the information from the retina varies from person to person.

How the brain perceives the information it receives from the retina is based mostly on past experiences, and there are experiments that prove this. Therefore, if imagination takes place when the brain perceives the information, and the way the brain perceives information is based on past experience, then imagination is limited by experience.

That might have seemed like a stretch, but think about this. When you are a kid your imagination is free flowing, but when you get older and gain experience your imagination begins to dwindle.

How does this apply to inventing?
Most of the great inventions that end up turning over a large profit or changing the world result from an invention idea that is completely ludicrous to the average person. This means the key ingredients to a successful inventor are a free-flowing imagination and the ability to creatively solve a problem.
In today's inventing world, many inventors have a background in engineering. I have been in engineering classes where they tried to teach me "how to invent".
They did it by rolling out 12 steps with each step instructing you exactly how to proceed and think. Did it work? No, of course not.

How to break down the barriers on imagination
Because the brain perceives information based on past experiences, a great way to train your brain to make new connections is by doing new things – by going to new places and meeting new people.

When your brain tries to fit these new experiences into the old familiar categories, it will find it easier to simply make new categories.

To change the way you think, it is helpful to look at things in a different way. Shift your perspective. Allow yourself to imagine different scenarios in everyday life – the child cleaning up after the parent – the dog walking the owner – the bird chasing the cat – the boss asking permission from the janitor.
Look at things differently. Put sunglasses on at night. Use a magnifying glass to look at an insect. Tilt your head sideways. Use two mirrors to see yourself as others see you.

Evaluating Your Invention Idea

If you are interested in generating income with your invention idea, then you aren't done. If you want to make money, you are going to have to spend a considerable amount of money to patent and possibly market your invention. You need to be certain you are completely happy with your invention and that it will be profitable when it is in the market. This requires more research. I know you are probably sick of research by now, but it's important.

There are three main ways to evaluate your invention.

1. Study past inventions. Look at successful past inventions and look at failed ones. Study how the economy was when the invention was released and how the economy compares to today. History is the best prediction for your success or failure.

2. Ask people you know. Talk to people about your idea and get their take on it. Talk to people that will be honest with you; false support can lead you to make a bad decision. Don't worry about them stealing your idea; most people are way too lazy to attempt that.

3. There are inventor organizations designed to help out inventors with ideas.

If you complete all of these steps with enough precision, you will have an invention idea worth patenting.

Invention Marketing & Licensing for the Inventor

Before you take any steps to market your invention, you should take a few preliminary steps.

Preliminary Patent Search

A preliminary patent search is generally a good first step. A preliminary search of various patent offices can be conducted for a reasonable fee (just contact a patent agent/lawyer), and it is even possible to conduct one for free (see the US patent office at http://www.uspto.gov/)

Patent Application

Don't publicly disclose your invention until after a patent application is filed. Publicly disclosing the invention before filing a patent application can potentially ruin the chances of ever being granted a valid patent. In fact, many Companies will not even talk to you until you have filed a patent application.

Prepare a Formal Information Package

You should prepare an informative and concise information package describing you, your invention and the potential market your invention reaches. The package should include color photographs of the invention, and a one page executive summary.

Prototype

It is a lot easier to sell a product if potential buyers can see, touch and feel the product. Building a working prototype is often a key step in selling your invention. Of course, some products are difficult to prototype, in which case a non-working mock-up may have to do. In any event, create the most professional prototype or mock-up you can.

Obtain Financing

Building prototypes and filing patent applications require funds. Finding that initial start-up funding is often difficult; however, there are two tried and true methods, namely partnerships and incorporations.

A signed partnership agreement is one way for a few people to pool their financial resources into a project.

If several investors are involved, then an incorporated company is a better method. Essentially, the company takes ownership of the invention and the investors contribute money to the company in exchange for shares. The number and price of the shares can be tailored to suit the particular needs of the project.

Now that we have dealt with some of the preliminary issues, let us look at the mechanics of selling your invention to a company.

The actual steps in the process are as follows:

1. Compiling a List of Potential Buyers

Finding a company that is willing to buy the invention is the most challenging part of the process. It begins by generating a list of companies that may be interested in the invention.

You can use a business directory to generate that list. Business directories list companies by the products they manufacture (or services they provide) and include basic information about these companies such as their address, phone and fax, and the name of the president (CEO or owner). Suitable business directories may be found in the business section of the local reference library.

2. Contacting Potential Buyers

Your list of potential buyers may include literally hundreds of companies. You simply call up each company on the list and ask them if they would be interested in receiving a solicitation for a new invention. Then get the contact information about the person you should send your information to.

3. Presenting the Invention to Prospects

After you have thinned out your list, your next step is to submit your information to each of the companies on the list. This may involve calling the people identified to be the "contact" for new product ideas and telling them that you are sending them an information package about your product.

Your package should include a cover letter and a one page synopsis of your product (including a picture). The information must be clear, concise and it must appear as professional as possible. Don't try to overwhelm the recipient - you want to impress them, not burden them.

4. Follow Up

Do not expect the prospect to come to a quick decision concerning the invention. It may take a prospect many months (even a year or more) to make up his/her mind on a project. You have to be patient. It is important to periodically follow up with the company but do not "pester" the prospect. Remember, the people considering your invention are probably quite busy with several other projects - annoying them may do little to speed the project up and may cause them to drop the project altogether.

5. Negotiations
If you find a company that is interested in picking up the project, then be ready to negotiate the terms of the sale. The key here is to be reasonable. From my experience, nothing kills off a potential licensing deal faster than an unreasonable inventor.

Realistically, the most you are likely to get is a good return on your investment. Asking for a smaller signing fee and a modest royalty is far more likely to generate a signed agreement than holding out for a big payoff.

6. Royalty Amount
I am usually asked the question "how much can I sell my invention for". I don't know the answer; however, here are a few rules which can help you figure out a reasonable royalty rate.

First of all, try to negotiate a royalty which is broken down into two parts, an initial signing payment and an annual royalty payment.

The initial payment should cover most of your costs of the project. The annual royalties should represent an amount which is sufficient to represent a good return on your investment without being a burden on the manufacturer.

The general "rule of thumb" is to ask for a small percentage (1% to 5%) of the net sales of the product. It is also possible, and in some cases advisable, to fix the annual royalty payment to an easily calculated amount (e.g. $1.00 per unit sold).

Selling your invention to a manufacturer is possible but it is not easy. How successful are you likely to be? From my experience, individual inventors are far more likely to successfully sell their invention by themselves then by going through some invention promotion organization. Like any business, the chances of success are a function of your determination, knowledge and willingness to take risks.
Article Source: http://www.gotinvention.com

Patent My Idea

What is a patent?
A patent is a licence granted to an inventor for a specific time. A patent protects a new idea by giving the individual the right to prevent others from making, importing or selling the invention without prior permission.

So how do I get a patent?
The first step is to conduct a worldwide patent search to see if the idea has already been patented. You do not want to waste many months waiting for the patent application only to be told that the idea has been patented by someone else.

What is a patent search?
It is a search carried out to establish if an idea or invention has already been patented. To be taken seriously by industry this needs to done by an expert patent researcher who carries out a thorough worldwide search using specialised techniques such as classification code systems and patent cross referencing. This patent search uses worldwide databases which are not accessible to the individual inventor.

Why have a professional patent search conducted?

- An advanced patent search can save the inventor a large amount of time and money at this stage of the process if a match is found and a patent cannot be obtained. It is better to establish this as soon as possible to minimize time and money spent on an idea that is not original.

- The idea may be an improvement on existing 'prior art' which could still be patentable. Therefore you would need to do the search to find all similar patents and then draft the application accordingly.
- A complete report is provided showing that a professional patent search has taken place on the specified invention. The report will provide the results from the search which can be used for the next stage of the invention process. i.e. you now know what areas of your idea need changing if they conflict with existing patents or which areas to focus on and further develop as they are truly original.
- Potential investors will expect a patent search to have been conducted.
- If the invention is found to be original through the patent search, it will help the success of the patent application, which is vital at this stage for the success of the idea.

Interpreting the results of a patent search

The results will always be useful either if the invention was found to be original or not. Results can be used for re-working and developing the invention. However the language used in patents can be difficult to understand and it is often hard to understand the implications of the results. Therefore it is important that any patent search that you commission comes with a discussion with an expert in the field of patenting and inventions to discuss the results.

Inventors need to be aware that there is no guarantee that when a patent search takes place that all prior inventions will be detectable. The patent office has a large amount of processing applications which have not yet been entered on the database.

It may also be that an application has been filed but not yet published. However it is still an essential stage.
Can I now patent my idea?
Now that a patent search has been conducted the inventor might choose to submit a patent application to the patent office. However it is likely that having looked at the results of the patent search some development needs to be done on the idea. This might be adapting the idea so that it does not conflict with any other patents or further developing the truly original aspects of the idea.

The Patent application
The patent application is the document submitted to the patent office to acquire a patent on an idea. Now the patent search has been completed and the idea has been developed, an application can be submitted to the patent office. This can be self-filed or done by a professional patent agent.

Application forms are available from the US Patent Office. It is an important document and so it may prove worthwhile to submit it through a professional patent attorney.

How to Patent an Idea
Patenting Will Increase Your Idea's Value

If you have come up with a brilliant idea that could be used within a manufacturing industry or business, the first thing you should do is patent that idea. The process of patenting will protect you.

If you do not patent your idea and discuss it with a company, and then in time you find out they have used your idea, you will have absolutely no recourse if you wanted to sue that company.

To know how to patent an idea, your invention should fall into one of the three categories below.

In the USA, a government-issued patent lets an individual stop other people from using or selling their item within the US, or import it into the US.

You cannot receive a patent on something that anyone could have figured out, or something like the law of gravity (it wasn't your idea!) or any printed materials (these are covered by copyright).
A Utility Patent protects the rights of an individual who has invented an innovative technological product, for example a machine, a chemical compound or a new component part of a machine. That patent will be extant for 20 years from the application date and a utility patent is the most commonly applied for patent.

A Design Patent protects a product's original ornamental design, but does not cover the design philosophy or mechanical characteristics. These patents last for 14 years from the date the patent was granted.

A Plant Patent is these days the least applied for and is issued when a new species of plant is discovered and this plant must be very different from previous discoveries.

There are other considerations in knowing how to patent your idea, and if it will be accepted.

Your invention must be useful (if it wasn't nobody would want it anyway!)

The idea should be a viable technical or industrial process, an innovative way of doing business or a new chemical mixture or compound that could be useful within a manufacturing process.

- It must be proven to work! That goes without explanation.
- It must be unique, not a new spin on something very similar - that may be an infringement.
- You cannot patent a basic or very simple idea; it has to be an item or process for which the inventor will have to submit a detailed description and drawings which will be scrutinized.
- You cannot patent earthquakes, fire, rainstorms or thunder, for obvious reasons.

To protect yourself from someone else using your idea you must patent that idea to receive the law's full protection and knowing how to patent an idea is very important.

You can later sell the patent rights outright, or enter into a licensing agreement with a manufacturer, which leaves you as owner of those rights.

If you have an idea that you think will make you millions of dollars, be prepared to hire a patent attorney, which will cost you a couple of thousand dollars. Patent infringement must be considered, and even after thorough research you may not be aware of a similar idea that has already been patented - so don't open yourself up to an expensive lawsuit!

When you know how to patent your idea successfully, be aware that it can cost you many thousands of dollars. But if you are onto a winner, your returns will be substantial.
Article Source: http://www.gotinvention.com

Easy Invention Ideas - How to Have Them

Want easy invention ideas? Dreaming up new products and inventions is fun, and it can be easy too.

Try the following two techniques and soon you'll have a list of new ideas.

<u>Start With What's There</u>
One of the easiest ways to create new ideas is to look at what already exists and find a way to make it better. You can start with things in your own home. These may even be the most marketable ideas - consider how many new kitchen gadgets are sold every year.

Look at the toaster, for example. How could it be improved, replaced, or the need for it eliminated?

You could eliminate the need for it if you designed a stove with a toaster built into it.

You could replace it with something like a waffle iron.

You could improve it by making it faster, perhaps with a combination toasting element and microwave heater.
Look around the room and pick out any item you see. Imagine how it would be if it was bigger, smaller, faster, slower, or different in some way.

As I write this, I am looking at a calculator. I would like to be able to talk to it. I would like to be able to say, "Mortgage payment, $140,000 loan, fifteen-year amortization, six point five percent interest rate," and have it tell me, "$1,219.56 per month."

With all the voice-recognition technology out there, this is possible. It just takes someone to figure it out.

Want an easy way to create new invention ideas fast?

- Make a list of everything in your house.
- Work your way down the list, thinking of some way to improve or re-invent each item.
- If nothing comes to mind, move on to the next item.

<u>Start With What Irritates</u>

Do you hate the way the ice builds up on the edge of your roof?

Do you get annoyed with the way the dog slops his water and food all over the kitchen floor?
Annoyances and irritations things are not just problems; they're excellent opportunities for easy invention ideas.

Suppose you are tired of burning your tongue on hot coffee. What could save you from this irritation? Perhaps a cup with a built-in thermometer that shows green once the coffee has cooled enough? Maybe a cooling device to set a coffee cup in, like a small fan that blows across the coffee when the cup is set on the device?

Annoyed with the necessity of brushing your teeth so often?

Maybe there is a Teflon-like coating that could be applied, so food wouldn't stick. If it was anti-bacterial as well, you might avoid plaque even without brushing.

Looking at what is around you and imagining small or large improvements is easy. It isn't difficult to train yourself to look at problems as opportunities and there are dozens of other techniques that will give you easy invention ideas. For now, start with these two simple ones and you can have a hundred new ideas today.
Article Source: http://www.gotinvention.com/detail.php

The Two Biggest Things Stopping You From Being A Successful Inventor

The saddest thing that can happen in life is unfulfilled potential. Tragically it happens all the time.

There are actually two main reasons for this.
a) Lack of information on Invention
It seems that most people have no idea what they need to do to develop an idea into a fully-fledged invention and then to bring that invention into the marketplace. There is plenty of ignorance where this subject is concerned and most folks believe that a fortune is required. The result is that many people give up long before they have even started and assume that because they are unable to raise the funds, the whole thing will not work out.

They forget that many inventions today were launched without any major funds or investment. Many were launched by very poor, poverty-stricken inventors and on many occasions the success of those inventions ended up turning around their lives.

Frank Stapleton's rather detailed ebook on invention can be very useful on this subject and will give you all the information required to go forward with your invention. Those who may not be able to afford it can take a look at this inventor's blog.

b) Fear of Invention Failure
One of the biggest obstacles that stands in the path of would-be inventors is the fear of failure. Many people are terrified of

making a fool of themselves. Some folks break into a sweat just imagining what a close relative of friend will think.

One of the basic rules for success is that the ladder of success can never be climbed with cold feet of fear.
Anybody who wants to be a successful inventor must overcome this fear and instead focus on their ideas and what they are trying to achieve. People will always have a good laugh at what you are trying to do, they have always done this, and they will always do it, so you might as well put all fear out of your mind.

Crowd-Funding

The other day, through random surfing around on the Internet, I discovered that there is a growing online trend: the concept of crowd-funding.

Perhaps it is not such a new concept, as politicians have been doing it for years to raise funds and even charities and entrepreneurs do it as well.

What is new, at least for me, is that it derives from crowd-sourcing or user-generated content. This allows work – or in this case, investing – to be outsourced to the masses or particular groups of individuals. In this way, crowd-funding becomes one part social networking and one part capital accumulation or fund-raising. What makes this even more interesting is that it is being used as a business model.

Essentially the model includes many micropayments that are made via supporters, or micro-investors, who all pledge to invest a small amount into a concept or product that will eventually pay off if enough people provide the determined amount and the product or service is produced.

If the determined amount is not reached there is no payoff to anyone. This can make it a bit of a gamble, unless the money is held in trust until the determined amount is reached.

Should the amount be reached, all the investors are rewarded in some manner, either in a cut of the total future sales or with goods or services exceeding their initial payment.

Crowd-funding entities seek to harness the enthusiasm and the cash of strangers, usually from the Internet, by promising them a cut of the returns (the incentive). In my opinion, this is similar to concepts found in gambling where a higher risk usually equals a higher payoff.

Although this concept is not particularly new, it is a recent arrival to the web and has quickly become quite mainstream and socially acceptable.

With popularity growing, it is likely that we will see a lot more businesses adopting this model, because:

1. The Web connects more and more people on a daily basis across the globe.
2. Increasingly, people are people willing to conduct financial transactions online.
3. The Global recession has made people more cautious with their investments. While this sounds counter-productive, the opposite is true. People are more likely to diversify their investments across many portfolios that ask for small payments, a trend which lends itself well to crowd-funding.

In conclusion, crowd-funding is set to grow rapidly as it is seen as a simpler and potentially faster method for raising awareness and/or money for a new invention. While it may not be an ideal model for every business or initiative, it has a definite appeal for people who are inspired by the pioneer spirit and the opportunity to be part of a new, emerging trend.

Currently it is primarily based on fund-raising and is focused on the idea of a community in which each individual can become a participant and part owner of ideas, goods or services.

Crowd-funding can be applied to any industry; as long as there is a community and a need, there will be all sorts of crowd-funding initiatives appearing. Take a look at the following examples:

IndieGoGo
Founded in 2008, indiegogo.com is a fund-raising website whose aim is to utilise an existing community to help people with any project, which can be anything from a cause, a film or a product that must reach a threshold by a defined timeframe.

IndieGoGo charges 4% of the total raised amount if you reach your threshold or 9% if the threshold is not reached.

Develop an Invention Business Plan for Success

Overview
An effective Invention Business Plan is an inventor's best tool for efficiently navigating through the invention process.

As an experienced inventor, I've learned that an idea is not perceived as a viable business opportunity until it can be effectively communicated on paper (or any other readable format).

No matter how great your idea is, most people expect that you have it written down. When I was new to inventing, I had no idea what that meant. I searched around but didn't find any universal format for documenting my ideas.

When submitting my concepts to invention hunts, licensing agents, manufacturers, retailers, engineers, and the patent office, I was asked many different kinds of questions.

The questions ranged from "What problem does it solve?" to "Who is your target market?"

Fortunately, with my entrepreneurial background and experience writing business proposals, I was very familiar with such questions.
To save time, I decided to consolidate them into a universal format that could be used and/or adapted for any audience within the invention process.

In this article, I discuss how to develop a versatile yet compelling business plan for inventors and their inventions.

I explain its importance, main elements, how and where to find content, and its many uses.

I also provide real examples adapted for three common purposes: for filing a provisional patent, for entering into an invention hunt, and for submitting to other key users.

Other key users may include retailers, manufacturers, industrial engineers, investors and licensing agents.

By sharing my insights and examples, I hope to help inventors like you develop your own material in order to effectively communicate and present your invention to the many different users within the invention process.

The Importance
An Invention Business Plan is an effective communication tool that provides a clear and tangible description of your invention while conveying its viability and value.

It tells a detailed story about your invention including what it is, how it works and why your invention is a believable business opportunity.

It can generally be described as an organized all-in-one depository of everything you know or have learned about your invention and it includes every detail about your invention so as to be used as a reference point for the development and/or submission of audience specific requests.

Having a broad audience scope allows it to be used as a collection of information which can then be modified or adjusted according to the audience in which it serves.

Invention Business Plan Example: The Main Elements
Many different readers and audiences need to see your idea in writing and you will be surprised how many different questions will be asked about your invention.

To efficiently answer such questions, the document should be designed to serve as a detailed yet practical guide and resource to be used by a broad audience.

Thus, the elements and content of your plan should be both comprehensive (i.e. can answer most questions about your invention) and adaptable (i.e. can be easily modified) for the purpose of a specific use or audience.

The recommended elements for a comprehensive and adaptable report are as follows:

- Short Description:
- A brief summary (1-3 sentences) describing your invention (name), what it does, and how it is useful.
- Abstract:
- A general description of your invention, its market, and its benefits. Include the target market, how your invention solves a problem, or how it is useful to your market.
- Fit
- How does your invention fit into an existing retailer or manufacturer's product mix?

- How does it innovate compared to their products? What is the best aisle to place your product?
- If possible, include a photo of the aisle and exact location on a shelf.
- List key selling and consumer advantages in a bulleted format. For example, key selling advantages may include up sell potential, a shelf attention getter, innovative disruptive qualities, and/or fills an underserved market niche.
- Consumer advantages may include simplicity, convenience of use, automates a manual task, saves time and steps, and/or solves an existing unmet need.
- Detailed Description
- This is where you describe the main parts or components that make up your invention, how your invention works or what it does, its main features, and method or intention of use.
- An example of main parts may include a container with lid, a motor for spinning, etc.).
- Examples of main features may include dishwasher safe, automated functionality, ease of use, etc.
- And, method of use examples could be: step 1, press red button to turn on, or pull white knob to make it move.
- Suggested Retail Price
- Base the suggested retail price on comparable market prices and other relative assumptions and factors.
- For example, if the invention combines the task of two or more existing products on the market, provide the cost of using those products separately and then demonstrate how your invention is priced such that it saves the consumer time and money.
- A good example is a food processor.

- You would provide the cost of knives, cutting boards, and the time it takes to cut everything. Whereas your invention, the food processor, is priced less than all of those things combined, plus you have the added value of convenience and time savings.
- Estimated Manufacturing Cost
- The ideal situation is to contact manufacturers to get a price quote of how much it would cost to build your invention. But this can be difficult if you don't have exact specifications.
- The other suggested general rule is to divide your Suggested Retail Price by a factor of 4. For example, if your suggested retail price is $80, then your Estimated Manufacturing Cost is $20.
- Problem/Challenge It Solves
- Discuss the details about the problem or challenge your invention solves.
- Include market trends and actual facts taken from credible sources.
- Describe how your invention is better than existing products.
- What are the flaws or downfalls of existing products and how does your invention solve those problems?
- Using the food processor invention as an example, you would say now it takes 20 minutes to cut vegetables for dinner using traditional methods (knives and cutting board). The food processor would reduce that time to 2 minutes.
- What Makes It Innovative
- How does your invention stand-out or how is it better than existing products or traditional methods?

- For example, the food processor saves users time, money, steps, and kitchen clutter in the food preparation process. Since there is no need to use multiple knives and cutting boards for cutting vegetables for dinner, you save cleanup time and counter space. Instead, consumers get a compact easy to use device with an automated motor for cutting vegetables to a desired size.
- Competition
- List existing comparable products or alternative methods currently sold or used on the market. Explain how your invention has a competitive advantage over these existing alternatives.
- Market Position or Target Market
- Who are the target users and/or target buyers?
- Who are the target retailers or manufacturers?
- What are the primary distribution channels (online, brick and mortar stores, both)? List examples.
- Packaging Suggestion
- How do you want to package your product such that it grabs the attention of the target user/buyer?
- Will your product be included as part of a kit of other products, or will it be a stand-alone product? Will it be in a box, a bag, with/without a label?
- What is your suggested package design and message?
- For inspiration, study package designs by other retailers or manufacturers.
- Product Extensions, Variations, Add-On Suggestions
- What other colors, designs, or styles can your invention have?
- Can there be multiple versions of your invention such as low-end and high-end versions?

- Can you add anything to your invention to make it even more useful?
- Do you want to provide a warranty for your invention?
- Intellectual Property
- Provide a patent number or provisional patent number if you have one.
- List the date and how you came up with the invention.
- Use the United States Patent Office website to research other related prior art.
- List and describe those related prior art.
- List the history of the invention, if any.
- Find descriptions of the history on any of prior art examples.
- List its primary components, provisional claims, and provide drawings or schematics of its design.
- Use prior art examples as your guide.
- You may decide to hire an industrial engineer, in which case, include those designs here.

How and Where to Find Content

While most of the content should be in your own words, the top five content resources for finding inspiration and ideas related to the above elements include:

- A related retailer or manufacturer's 10-Q (Quarterly Statement) or 10-K (Annual Statement)
- Look for Market Research, Problem/Challenge It Solves, Competition, and Target Users.
- 10-Q's and 10-K's can be found on the website of most public companies. If not, search the Security and Exchange Commission's (SEC.gov) website.
- Related patents from the United States Patent and Trademark Office (USPTO.gov)

- Look for Abstract, Detailed Description, Components, Features, Methods, Intellectual Property Research, Patent Results, and Drawings and Design.
- Trade association websites, magazines, and other trade materials
- Look for Market Research, Competition, and Target Users. For example, a well-known trade association is the International Housewares Association (IHA).
- Websites of retailers or manufacturers
- Look for Packaging Suggestion, Product Variations, Key Selling Advantages, Key Consumer Advantages, Suggested Retail Price, Manufacturing Cost (general rule: divide retail price by 4).

Usefulness and Audience

The invention process involves disclosing your invention to a wide variety of readers. Each document is a starting point or template for providing future material with respect to the many different readers and audiences for which you will need to communicate your invention.

These audience may include, invention hunts, industrial designers or engineers, retailers, manufacturers, licensing agents, lawyers, marketing agencies, and the United States Patent Office.

With a well-documented plan, you can conveniently adapt or modify it, depending on its primary use or audience, thereby saving you time and steps. As a general rule, be conservative about what you disclose. Only provide information that is requested or required.

I also suggest including a non-disclosure agreement (even if a provisional patent is filed).

Conclusion

An Invention Business Plan is an essential part of the invention process. It helps inventors effectively convey an idea into a tangible, understandable, and justifiable business opportunity. I wish you much success with your invention endeavors.
http://gotinvention.com/detail.php

Small Business and Social Media

[Brian's note: I really enjoyed this article. Even though it was written in 2013, it still hits home when the author discusses the importance of Social Media at all levels of business, especially for inventors just getting started who want to get the word out about their big ideas!]

Posted on May 29, 2013 by Chris in Social Media

In an increasingly competitive market, business owners are constantly on the lookout for a way to improve on current sales figures, boost the current rate of brand growth and beat out their competitors.

This is all good and well if you have significant cash flow but what small businesses often restricted by a minimal task force, severe time restrictions, and near nonexistent budgets?

The Big and The Small
There are very clear distinctions between the large corporations and the small businesses. One has the time, resources and money to grow whereas the other simply doesn't. But, and it is a big but, there is one method of marketing that requires minimal time and effort that can potentially provide a much needed boost.

I am referring to social media marketing.

As I say this, I do not doubt that there are many small business owners rolling their eyes and saying they simply do not have the time to waste on such things.

As recreational as many of them may seem; these sites have changed the way we communicate. As of 2014, Facebook had more than 1.25 billion active members and Twitter members sent more than 500 million tweets per day. On YouTube, 6 billion hours of video were watched each month. LinkedIn, Google+ and Instagram each had 300 million monthly users and Pinterest had 70 million users (80% women) each month.

These sites are used multiple times a day by their members and are now the first place consumers air their disdain or happiness over a product and service.

When you realize that these sites are the first place people look for recommendations and advice… Social Media doesn't seem to be such a big waste of time.

Social Media and Consumerism

Social media sites have an increasing influence on consumer choice. This is why, whatever the business size and whatever the industry, it is foolish not to have a presence within the social media marketplace.

For the small business owner, social media marketing may appear to be a waste of time. How can you possibly compete with Coca Cola and its sixty five million fans? Or with BBC News and the constant feeds it provides to its seven hundred and eight five thousand followers?

The bottom line is that you don't have to. The small business owner does not have to think big to achieve success. You simply have to know your audience and offer them just what they need. That's all there is to it.

Do a little market research to find the answers you will need before you begin your journey:

- Which social media sites does my target audience use?
- What does my audience look for on social media?
- Which groups is my audience likely to follow? Why?
- Is my audience regularly active on their social profiles?

By determining the answers to the above you can figure out a solid social strategy…

- A Facebook page that shares engaging content, interesting images and offers crucial advice and support
- A Twitter page that offers exclusive deals, the latest pricing information and any discounts and offers
- A YouTube channel that is informative and engaging
- A Google+, LinkedIn or Instagram page offering relevant tools, assistance and guidance.

Any page/profile on any social media site takes just a few minutes to set up, is free of charge and easy to use.

No matter how restricted your time may be, as a small business owner you can benefit greatly by investing the time to create a presence on a social media site and then link it to your website, email signatures and any other online marketing tools you use.

From there, encourage your staff, family members and friends to Like and Follow you on these sites to start building a community of supporters.

After that, make sure you set aside at least 10 minutes a day to market your business socially by commenting on current events, including:

- Industry news
- New stories
- Facts
- Advice and support guides
- Business/staff related news
- Local news
- Blog posts
- Engaging imagery and videos
- Product/Service offers and deals

By actively offering some or all of the above to your social media followers, you stand a better chance of ensuring that your profile achieves active engagement.

In social media, it's all about offering something worthy and interesting, so before you post, ask yourself: will my target audience like this? Will they share it and tell their friends about it?

Social Media and Search Engines

Google is a search tool that tracks and records your activities on the internet. It notices which sites you frequent and takes your habits into account when it delivers organic search results. Facebook and Bing work in a similar fashion.

How this serves you is that your business can effortlessly improve its organic search results across two the internet's major search engines simply because you take the time to visit your social media pages.

Social Targets

Once you establish clear guidelines, just set your business a small target… it could be to receive a certain number of likes, shares and re-tweets a week or maybe one on one engagement with a potential or existing customer. Whatever it is and whatever you do, as long as you are active and realistic in your approach it doesn't matter what size your business is, social media marketing can help give you a much needed boost.

WHAT OTHER'S SAY ABOUT BRIAN FRIED

William Jones

Supervisor Radiation Oncology at Cancer Treatment Centers of America

Brian promptly called me. We had a productive first conversation in regard to my three potential inventions. Brian was kind, knowledgeable and willing to follow-up with me if necessary. Excellent interaction. I look forward to working with him in the future.

Cindita Cunningham

Owner at Cunningham-Innovations

Brian is intelligent, has a positive attitude, teamwork mentality and excellent communication skills. He is easy to work with and trustworthy. Throughout the time my husband and I have known him he has shown himself to be a motivated leader and amazing mentor

Olga A.

If you are an inventor and don't know who to call and where to begin Brian is the answer. Extremely trustworthy and knowledgeable, he will help you decide if taking your idea to the next level is the answer for you. His personal knowledge of patents, licensing deals and source of manufacturers is refreshing. I would recommend Brian in a heartbeat.

Jessica Dighton

Co-Founder/President Lickety Klip® - New Multi-Solution Accessory

Brian was exactly the person we needed. After hearing our position, knowing our wants and desires and weighing all options for our business, Brian helped us strategize a plan that was best suited for us and our business. It's comforting getting counsel from an expert who has the inventor's best interest in mind. Thank you again Brian!

Greg Ryan

It wasn't until this past year, when I met Brian Fried, after one of the Inventors and Entrepreneurs Club Meetings, that I'm finally experiencing continued success with one of my inventions. I know Brian to be an honest, sincere person, and would recommend anyone looking for help with their Invention Business.

Brian A. Donnelly

Brian Fried is an inventor's consultant "for all seasons." His knowledge, experience, professionalism, and humanity have been very helpful in advancing my invention. Moreover, he helps other inventors by imparting useful information and advice as founder and director of the

Suffolk County Inventors and Entrepreneurs Club. Also, his internet radio show <Got Invention.com> is a treasure trove of useful information for inventors.

Freeman Fields I

It was an honor being able to listen to Brian speak at my inventor's group. I was truly amazed at his level of expertise with product licensing, as well as his complete commitment to educate inventors and advance their concepts. Although i have been working with patents & inventions for over 20 yrs, i was able to gleam so much knowledge from just one session with Brian. With that being said, i truly look forward to working with him in the near future. Thanks for caring

Maria Ruvio

Consumer Product Developer

Brian Fried is an entrepreneur with passion in innovation. In the past 5 years that I've know Brian, he has continuously displayed dedication and education to the world of inventing and inventors alike. His contribution range from leading local inventors group, to radio shows and even has appeared on national TV. His passion and contributions to the inventor's world continues to evolve and the best is yet to come from Brian.

Richard Alpert

I recommend Brian without reservation. In the vernacular, he is "the real deal". In my interaction with Brian, I have been impressed by his knowledge and his integrity. He is sincerely interesting in helping his fellow inventor, illuminating the path to successful market research, patenting, and production and distribution or licensing, all the while pointing out possible pitfalls. He's a good person to know.

Hilaire Toto

Vive La Vie, Soup for Life!

In spite of a very busy schedule and an extended appointments, Brian made himself available to me when I connected with him. Brian is very courteous and business savvy. His recommendations and insight proved to be a game changer.

Thank you Brian for your time and availability

Nicholas Caracappa

If you are an inventor who needs help, Brian is here for you. He has the business experience and connections in the very circumscribed field of inventions and intellectual property. Being a successful inventor himself, he has the knowledge and experience to help you take your invention to market. There are few who can invent and even fewer who can effectively turn a patent into a profit. Brian is one who will help you do just that.

Colleen L. Costello

Product Development Engineer I Speaker, Writer, and Educator

Brian worked with me as an inventing mentor. During our discussions, Brian was insightful, respectful, direct and easy to talk with. His knowledge of the independent inventing industry is vast and he readily shared that knowledge with me without an air of superiority always respecting my work and opinion. He offers a balanced approach in his advice since he has both licensed his own products and brought them to market himself. Lastly, at the end of the conversation, he helped me formulate a clear plan of action with do-able action items. I would recommend Brian to any inventor.

Madeleine Schwartz

CEO of I START HERE ALWAYS, LTD.

Brian is extremely knowledgeable in the field of marketing one's invention. His grasp of comprehending the numerous marketing techniques gives his audiences and clients a perspective that most people do not get to see. He is willing to share what he knows so that his own experiences of winning and losing helps other inventors.

Timothy Mardis

IntelliHUB

I would like to take a moment and personally thank Mr. Brian Fried for all that he does for those of us in the invention community. I'm not really sure what it is that Brain takes to keep him going all-day, but this young man is involved in so many business affairs that it's difficult to conceive how he does it all.

If by chance you've ever heard of Inventor Chat or Invention Radio or Got Invention or You & Your Big

Ideas (I've probably overlooked something) then that is my friend Brian Fried. And although I've never had the pleasure to personally meet Brian, I feel as if though he is a friend for several reasons, but the one that I admire the most is that he works tirelessly to help others without hesitation.

Just today I sent him an email (you can find his contact information on any of his websites) and asked him to refer me to a Patent Search firm and within a matter of a minutes he responded with an answer. Come to think of it, I'm not sure any of those I've known for years would reply that quickly.

Do yourself a favor, take a moment and get to know Brain, purchase his eye-opening book "You & Your Big Ideas" and brainstorm with him... you'll be amazed by his wisdom and willingness to help.

Cornel Chin

Celebrity Fitness Consultant/Troubleshooter/Fitness Author & Inventor

I came across Brian's radio show Got Invention Radio by pure fluke during the embryonic stage of my fitness invention. Having listened to many of his shows via podcast, with Brian's many expert guests I managed to gain a very broad knowledge of how the complicated invention world worked. This has not only saved me huge expenditure, but as importantly valuable time. Brian has also taught me the absolute importance of checking for prior art of a proposed invention/idea. He's an incredibly busy person, but has a real passion for what he does and how he helps others and I am truly grateful to him for this.

Terry Whipple

Founder/Facilitator at Inventors & Entrepreneurs Club

Brian is a person all inventors needs to know. Not only does he have a passion to see others succeed, he is also a great resource for those that need a help in exploring and capitalizing on their idea. Brian has a network of valuable contacts and has shared it with many of the Inventors & Entrepreneurs in our Club. Thanks Brian!!

Peter Janora / PJMAX Solution

Brian Fried is a Big Idea Guy! He has the knowledge and experience to assist the inventor in the invention process and has navigated this process successfully. His book You and Your Big Ideas is a great resource for the inventor and I recommend it to all inventors . Brian conducts his business and radio show Got Invention Radio with passion and professionalism. If you are looking to bring your Big Idea to the market, read his book, listen to the radio show and you will be inspired and to make your products and dreams a reality!

Jay Welikson

A year ago, I read a newspaper article about Brian Fried. It showcased his entrepreneurial skills as an inventor, a promotions genius, successful marketer, and is well connected politically. What really piqued my interest and trust was that he gives back to the community.

Every month, Brian conducts seminars for inventors and entrepreneurs who don't have the marketing skills or necessary connections to push their products to levels of distributorship. Brian gives of his own time to help these people with tangible advice. I find that quality about Brian very admirable.

Brian's business acumen and energy are of the highest level. Brian is quickly able to ascertain a product's viability and marketability. His business skills allow him to make dreams into reality. Such is the case with me.

I've been in the same industry for decades. Within that industry, I invented and patented a new category. While I have manufactured, procured and sold hundreds of millions of dollars of product, I was unable to successfully market my product due to capital limitations.

Brian quickly assessed the high potential of my product, believes in it wholeheartedly, and has moved quickly to make the necessary connections to market the product. I have made Brian a partner on this project on a handshake, and we've made amazing progress.

I have never had a partnership based on a handshake. My industry doesn't have room for trust, is notorious for copycats, and doesn't suffer fools gladly.

Within several days of meeting Brian, I came to trust him as I trust my family. I don't believe I can say enough good about Brian's integrity.

Gene Benfatti

Brian Fried is the shining light to guide you down the path of invention. He's a man of great integrity, knowledge and passion when you need advice from a fellow inventor who's always available to answer your questions, and help you any way he can. I Highly Recommend Brian as an Inventor's Consultant, Mentor, and Friend.

Suzan Koc

Songwriter's Rendez-Vous

Brian helped my mother, who is an inventor, through the licensing process as well as helping get the right people for the infomercial. His insight has been invaluable. We are very fortunate to have him on our side.